SIGINT

SIGINT

THE SECRET HISTORY OF SIGNALS INTELLIGENCE 1914–45

PETER MATTHEWS

To my grandchildren, Hannah, Alex, Piers,
Elliot, Natasha and one yet unborn.

First published 2013

The History Press
The Mill, Brimscombe Port
Stroud, Gloucestershire, GL5 2QG
www.thehistorypress.co.uk

British Library Cataloguing in Publication Data.
A catalogue record for this book is available from the British Library.

ISBN 978 0 7524 8734 2
Typesetting and origination by The History Press
Printed in Great Britain

Contents

Acknowledgements

I am pleased to acknowledge the advice and help of all those mentioned below and thank them for helping iron out some of the wrinkles in the book:

- James Gates, who helped with validating some coding exercises and assisting with the mysteries of computing when several chapters of my book disappeared from my computer's memory.
- Joseph Mankowitz, who encouraged me with snippets of military history and participated in an exciting visit to Chicksands Priory which was a Listening Station for the Royal Air Force during the war. It is still in operation as a British Army Intelligence Corps unit.
- Paul Croxson, who was my rock in writing the book and put me right on much in the intelligence world with kind comment and guidance. He served in the Intelligence Corps in GCHQ and Germany as a traffic analyst. He is now a voluntary archivist in the Intelligence Corps Museum at Chicksands Priory and researching the history of SIGINT.
- Marylyn Blackwell provided information about the *Tirpitz* from the archives of Lieutenant Norman Gryspeert RN.
- The British Library staff and, particularly, Cathy Collins who guided me towards the accounts of the wartime

actions in the four-page skimpy newspapers of the time in the new National Press Archive.

- Clive Matthews, who instructed me in the use of social media and helped me carry out an interview on the CNN Network in the USA about this book.
- The Public Records Office (National Archives, Kew) where many obliging staff guided me towards the captured logs of U-boats and British warships fighting the war at sea in both wars. The log of my grandfather's ship HMS *Venus* in operation from Queenstown in Southern Ireland was of particular interest to me.
- Various members of the Imperial War Museum in London, but in particular Nick Vanderpeet and his staff at its Formal Learning Department. The exhibits in the Secret War section of agents operating gave the 'feel' of the equipment they used.
- The German Bundesarchiv in Koblenz have guided me through the early history of the German Federal Intelligence Agency (*Bundesnachrichtendienst* or BND for short) where Dr Hechelhammer has helped with information about the spymaster General Gehlen.
- Randy Rezabek in Los Angeles, who maintains an encylopedic archive on the subject of TICOM (Target Intelligence Committee) and provided documentation on German intelligence matters, some of which are quoted in this book.
- Werner Sunkel, a volunteer curator at the Wehrtechnik Museum in Nurnberg, whose father was *Oberwachtmeister* (Sergeant Major) at Lauf and left photos, some of which are reproduced here.
- Dr Steven Weiss, with whom I had discussions about Operation Anvil, in which he was involved as a soldier, and his experiences with the French Resistance, and he kindly shared his photographs with me for this book.
- Members of the Tunbridge Wells Library who surprised me by seeking out many books for my research.
- The German Wehrmacht veterans of the Abwehr (German military intelligence), most of whose names I have forgotten, whom I knew and talked with in Berlin

as the Second World War ended and the Cold War began. Their wartime experiences, often recounted at great length during the long days and dark nights of the Berlin Airlift, are used in this book.

- Wilhelm H. Flicke (now deceased) who I got to know well was one of those mentioned above who gave me his papers that have formed a part of this book. I have had his badly typed memoirs deposited in the rare books section of the research library of the Imperial War Museum.
- Last and most appreciated is my wife Carole, who supported me uncomplainingly during the hours that I spent at the computer and even read and corrected the manuscript of this book.

Preface

As war ended in 1945, some of the German Wehrmacht's most experienced signals intelligence operators began working for Anglo-American intelligence against the Russians. As a young soldier I mixed with them, perfecting my German as I talked with them at length and listened to their experiences. I was tolerated because I brought cigarettes, which were in very short supply in Germany at that time. They believed themselves to be the elite in signals intelligence at that time – and, from their tales related to me in 1947, so did I. The triumphs and few tragedies they related at that time sounded good, but they knew little of the reverses that the Wehrmacht suffered in the signals intelligence war. So we all believed their tales of outwitting the enemy. It was not until almost thirty years later that the story of Bletchley Park began to emerge and I can imagine the astonishment they felt as the secrets began to be revealed. Most of the German signals intelligence warriors that I had met with and listened to with rapped attention in 1947 would have been dead by the time that we all knew of the story of Ultra intelligence. They have left me impressed with their achievements, even though their work had been eclipsed by Bletchley Park. They even gave me some of their private papers (bought from them for a few cigarettes which they probably used in trading on the black market).

I have been re-reading their accounts of the 1940s and have tried to set them into the considerable body of literature that has been published about British cryptography and smiled at the naivety of those German cryptographers. I decided to write this book to try and compare the abilities and shortcomings of the signals intelligence agencies of the belligerents in the two great wars of the twentieth century.

The book traces the emergence and maturity of electronic signals interception and decryption in the first half of the twentieth century, and its effect on the warring nations of the world. The place of signals intelligence, or SIGINT, in military history is not just about Bletchley Park and Enigma, but the way that the intercept services of armies, navies and air forces competed to gain advantage in battle. Intercept services of Germany's Abwehr, France's *Deuxième Bureau* and the British Admiralty Room 40 all vied with each other in their different fields from the inception of signals intelligence before the First World War. The first steps in electronic warfare had been laid and preparations were made during the inter-war years for the signals intelligence battle in the Second World War. This is its tumultuous story.

The author's interest in wireless telegraphy, or W/T as the Royal Navy called it, began in his home town of Portsmouth where as schoolboys we were obsessed with naval matters. We all learned the Morse code at an early age and understood the concept of its communication at great distances, which is more than some of our elders managed to do.

My generation received the call to arms just as the war had ended and the fighting ceased, but as we joined the army we were to get a master class in signals warfare. We were ordered to act as garrison in Berlin and flew there in an old Dakota aircraft, which was the first flight of an operation that came to be known as the Berlin Airlift. The Russians were laying siege to the city and all food, fuel and other supplies had to be flown in by a great armada of planes to supply over 2 million people. The Berlin Airlift was a gesture of defiance to Russian demands and marked the beginning of the Cold War.

Berlin was in ruins and the most despondent place I had ever been in my life. We mixed with the Russians and even shared quarters with Red Army soldiers during our duties in Spandau Prison, where we provided the guard for Hess, Raeder, Speer and a few other Nazi war criminals. Berlin was steeped in the espionage business; the city contained one of the most forward radio listening stations that the West had in the signals intelligence war that was developing with the Soviet Union. Listening posts scanned the electronic chatter of the Red Army, and the Allies often used ex-Wehrmacht radio veterans who had served in the German intercept service during the war. Their experience was being used and analysed by the Allies through TICOM (the Target Intelligence Committee project formed to collect information about German intelligence activities during the Second World War and seize materials) and the emerging Gehlen Organisation to try and read Russian signal traffic. My German friends and I often conversed during the cold nights over a mug of cocoa, listening to their reminiscences, some of which appear in this book. From our discussions I realised to what extent interception, decryption and deception was possible in signals operations and how that intelligence could be used by the enemy. The game of electronic hide and seek in the ether, and the way it could be played, fascinated me.

Later in civilian life it seemed to me that the country was obsessed with spies in the 1950s; Burgess, Maclean, Philby in the Foreign Office, and Blunt in the Queen's Household, were all found spying for Russia. I had seen enough of that scene in Berlin where every fourth person seemed to be a spy working in the pay of the Russians, the Americans or both. In spite of this, I began work in information research, concerning myself with how Britain handled the surveillance of German newspapers, magazines and other publications during the war. Before the war, publications were easily obtained, but with the outbreak of hostilities the supply dried up dramatically. Newspapers and other publications could only be smuggled out one or two copies at a time. The intelligence community demanded their copy to read as the contents often contained clues as to the state of the enemy's mind and resources. Microfilmed copies were circulated to a distribution list, compiled of various Ministries in Whitehall, and Army and Navy intelligence officers needing to read the enemy's newspapers. One of those on the list was Bletchley Park, but that did not mean anything to us then. Clues such as announcements about rationing regulations indicating the condition of food supplies in a region or an announcement of a German officer's marriage that might identify his regiment could be gleaned from their local or national newspapers. To satisfy the increasing demand for the newspapers, magazines and even scientific papers of German and occupied countries that had been 'obtained', the originals were distributed by despatch riders, then delivered to many obscure addresses all over the country.

Later on, our tenuous connection with Bletchley Park earned us a tour conducted in person by its director and ex-secret service man, the late Anthony Sale. Tea and biscuits were followed by a discussion about the intelligence scene. To our surprise, Sale told us there was relatively little in the archives about how enemy intercept services operated during the war, in spite of the TICOM investigation. I searched out the papers stored away, particularly those given to me by Wilhelm Flicke, a German cryptographer, relating some of his experiences over twenty-five years of cryptography in the service of his country. His experience started in 1917 and spanned the inter-war years and, finally, in the German signals intelligence service throughout the Second World War and even beyond. Contributions from Colonel Randeweg, who had commanded German intercept units on the Western Front, as well as from other signals operatives, have shaped this book.

The reader must remember that original papers and reminiscences of German signals operatives from the time are not informed by the knowledge we have now of the intelligence activities that were operating all over the world, and that, at the time, none of us knew of the existence of Bletchley Park. German Abwehr intelligence thought that they had performed better than they did because they did not know what the opposition was doing, so some of the views that are quoted may seem a bit quaint to an informed reader. That does not disguise the fact that the Abwehr did enjoy some successes in the signals intelligence war in which they were engaged. This book is a history of SIGINT operations largely from the Axis partners' point of view and how the intercept services perceived themselves. Its aim is to tell the other side of the Bletchley Park story.

Peter Matthews, 2013

Introduction to Signals Intelligence

The SIGINT Battlefield

This is not only the story of Bletchley Park and Ultra intelligence, and its incredibly clever and dedicated team which helped to win the war for the Allies. Less is known about Bletchley Park's opponents in signals intelligence and what contributions they made to the war effort of their forces in comparison to the Allies. The German Enigma and Lorenz, the Italian Hagelin and the Japanese Purple coding machines were all penetrated by Bletchley Park. The German Abwehr intelligence machine in particular made progress of its own into Allied codes and ciphers; the Wehrmacht (German armed forces) believed they were winning the signals intelligence war and so they were in the early stages of the war. The complex struggle for advantage in a battle of wits between intelligence agencies and the way it was resolved is a tragedy followed by a triumph for German intelligence. Its legacy still lives on in the world's intelligence community today, but its story began at the start of the last century.

A signals intelligence war is a most curious conflict; its participants very rarely meet and they will never know who has won the battle until years later, if ever. For instance, Colonel Walter Nicolai, head of the German intelligence agency during the First World War (1914–1916), believed that his agency dominated the intelligence scene in that war. Nicolai had a fearsome reputation as a spy master and had enormous faith in his own ability. What he did not know, however, was the extraordinary success that the British had in breaking the codes in the wireless messages his department sent, and how the British became the real victors in the war's signals intelligence struggle. A clearer picture only emerged in 1982 when Patrick Beesly published his book entitled *Room 40: British Naval Intelligence 1914–18*, spelling out the achievements of the British code breakers. Nicolai never knew the truth as he died in a Soviet prison just after the Second World War. Similarly, in the Second World War, when officers of the German Abwehr signals intelligence agency boasted of their efficiency and achievements in their memoirs, they knew nothing of their opponent's centre of excellence, Bletchley Park. The first book about Ultra intelligence derived from decoding the Enigma and other machines was published in 1976, many years after the war. Some of the German intelligence reports in the *Deutsches Bundesarchiv* (German Federal Archive) are amusingly naive in believing that they are the keepers of the military knowledge when Bletchley Park was looking over their shoulder and reading their messages all the time.

However, we should start at the beginning. Signals intelligence, or SIGINT as it was abbreviated by the Allies in the Second World War, is primarily an aspect of military intelligence gathering based on the interception and decoding of an enemy's wireless transmissions. SIGINT has played an important part in warfare for just over a hundred turbulent and bloody years, and proved a powerful weapon in the hands of many commanders in the field and admirals at sea. Intelligence of this kind has developed an array of scientific techniques for eavesdropping on opponents, which are now used by most of the world's armies, navies and air forces. It was developed during the huge conflicts that were fought in the first part of the last century when wireless messages became central to the mechanisation and mass movements of formations on the battlefield, as well as commerce raiders and submarines at sea. Intercepted signals come straight from the mouth of the enemy and could sometimes give an immediate indication of the strength, order of battle and possibly even objectives, both tactical and strategic, of an opponent in the field. Signals intelligence has proved to be a very cost-effective method of building a picture of the enemy's strength and intentions, compared with other aspects of intelligence, although other information sources cannot be neglected in the building of an intelligence evaluation.

Throughout history military messages have had the complication of being coded, but the interception and deciphering of wireless transmitted coded signals has been one of the intellectual adventure stories of modern military and even political history. The complexity of codes and ciphers are only outlined here, although some of the principles are discussed; however, it is the effects that SIGINT has had on the conduct of military, naval and air actions which are the subject of this book. Intelligence has been a significant weapon in both of the world wars and Britain's part as played out in the Admiralty's Room 40 during the First World War and particularly Bletchley Park in the Second World War are well documented. People may be forgiven for thinking that these were the only signals intelligence operations of any importance during the two great conflicts, but this book endeavours to show aspects of the enemy's signals intelligence offensive during those times. SIGINT achievements of the various combatants have ebbed and flowed over the years and their success, or lack of it, in the development of military intelligence has been largely dependent on their understanding of the concept. The function and application of intelligence by its commanders and their masters, and their ability to deploy this unique weapon of war, has not always been shown to the best advantage. A successful signals intelligence service needs to have a champion, or at least a tacit acceptance of its value in shaping battles at the highest political level. The attitude of the leaders at the highest political level to an intelligence function and, particularly, the value of the immediacy of those intercepted signals has been essential to the effectiveness of intelligence services and the military organisations that served them.

The Value of SIGINT

SIGINT was to become an important indicator of an enemy's, or even a neutral country's, intention to act. The decrypting and reading of an opponent's wireless transmissions, which might contain reports of actions or movement orders for troops or ships, could be an important part of that indicator. The deciphering of diplomatic wireless transmissions that send or receive instructions in negotiations has proved to be of great consequence. An operator transmitting his message is, by and large, confident that he is not going to be overheard or, at least, if his message is encoded then it will not be understood. To get into the mind of a potential opponent is as important to a diplomat as it can be to a soldier.

Enemy signals gathered for assessment and encryption had to find their clues where they could from the 'raw' interceptions their listening stations were receiving. Putting together the pieces of a signal, or more probably hundreds of signals with observations from other sources, helped to create a huge jigsaw puzzle in need of evaluation. However, an intelligence officer invariably did not have all the pieces and there was always the possibility that some pieces might belong to another puzzle. They might even have been put there maliciously by a cunning advisory to make the task of clarifying the message even more difficult. General Montgomery's intelligence officer, Sir Edgar Williams, expressed the view that military intelligence is always out of date as there was always a time lapse due to the speed of events. He went on to say that it was better that the best half truth was received on time rather than the whole truth too late. Speed in delivery of an evaluation is, therefore, often the essence of the matter. Many Ultra decrypted messages were in the hands of the commander that could use them within three hours of them being transmitted by the Germans.

SIGINT has been deployed in many forms in every theatre of war by combatants since its inception before the beginning of the last century, albeit with various levels of success. Ultra, which was Britain's great triumph in signals intelligence in the Second World War, is estimated by some to have shortened the war by up to two years. Other less well-documented SIGINT engagements also pose the question: 'By how much has SIGINT lengthened our wars?' Its successes and failures undoubtedly helped to do that in the first months of the First World War, as we shall see. Similarly, events and actions such as the interception of British Army plans in the Western desert of North Africa could have also had the effects of prolonging the war in this theatre. In the end, however, the pluses and minuses of the many actions influenced by signals intelligence are too complicated to calculate what effect they may have had on the duration of hostilities.

Intelligence reports do not win battles, but merely enable a competent commander to deploy his forces to better effect against his enemy if he knows the strength and deployment of the opposing forces and what objectives they intend to achieve. Conversely, good intelligence can be of little help to a poor field commander who cannot take the right decisions or does not have sufficient resources to fight his battle. The Battle for Crete, where General Freyberg knew from Ultra intercepts when and where German paratroops were to land but neglected to secure the runway at Maleme Airport with a few anti-aircraft obstacles, is a case in point. Denying the Luftwaffe a landing ground for troop-carrying aircraft with reinforcements may have influenced the tide of battle on the island. Battles are won by the good and well-prepared plans enacted by an able commander who has sufficient resources to fight a particular battle. Intelligence could be a powerful influence on these plans, or the commander's reactions in battle. As a result, signals intelligence has helped to determine the outcome of not just battles, but given shape to the conduct of campaigns and even a war.

1

From Cables to Codes

Cable Wars

Telephone and telegraph communication networks had already changed the world of commerce and diplomacy in the ten years before the war. Now the nature of war itself was changing beyond imagining, commanders in the field were enabled to control huge troop movements or direct artillery fire onto targets from remote control emplacements or even aircraft spotters. The technology enabling generals to direct operations on fronts hundreds of miles away created a form of warfare of unprecedented scale and proportions. To most of Europe's young adult men about to be called to join the service of their country at the beginning of the last century, being able to communicate over great distances had an almost magical feel. Germany's scientists were in the forefront of this new communications revolution and Imperial Germany's government saw great possibilities in the telephone well before the war began. An ambitiously large network of telegraphic wires had been laid across Germany connecting important cities and, in particular, their military establishments. The High Command would be able to direct the great battles yet to come, using their newly installed telegraphic communications network, making the industrial scale of the First World War a possibility. This new technology would have far reaching effects on the war's fast-moving opening stages.

As the First World War began, initial actions were directed against communications networks in the shape of the great undersea cables connecting the major cities of the world. Cables were far more important to international communications at that time than radio telegraphy. The earliest cables were laid in the 1850s and the first cable links between Europe and America, or rather Canada, were made at that time. Complex cable networks linked the colonies of both the British and the German empires to their capitals and homelands; these links had been laid years before to transmit market and financial information to support the international trade of the world.

On the other hand, wireless technology was still emerging, although Germany, using the Nauen wireless station near Berlin, had just made her first long-range wireless transmission in 1913 to her station in Togo, German West Africa. The transmission was barely audible and, by comparison with a cable message that would have been clear and distinct, it seemed that wireless telegraphy still had some way to go. Britain had extended her own cable networks to every important part of her empire as a priority and owned, or at least controlled, much of the world's networks. She was also experimenting with wireless telegraphy and was slowly increasing the distance of global wireless transmissions. In 1910, communication was even made with an aeroplane in flight. Cable links, however, would initially prove more vulnerable to attack than wireless so, two years before the war started, the Committee of Imperial Defence planned a crucial action that determined the shape of the future signals intelligence war. A secret 'war reserve' standing order was issued by the British government to cut Germany's international undersea cables in the event of hostilities.

Just two days after war was declared, the cable ship *Telconia* slipped her moorings in Harwich on 5 August 1914 to steam out to a position near Emden on the German coast to dredge up, lift and cut five German telegraphic cables. The cables that were severed ran down the English Channel to connect with France, neutral Spain, Africa and to Germany's many friends in North and South America. A few days later, with second thoughts, the *Telconia* returned to dredge up and reel in many more thousands of feet of cable to ensure that the cable damage was irreparable. The Imperial German government could no longer send urgent cablegrams to their African colonies or embassies in many neutral countries around the world. One of the war's first military actions by the British was to raid and destroy the German cable station in Lomé in West Africa, isolating German cruisers on station in South Atlantic waters. Without the cable-based telegraphic communication to their bases, they would have to use wireless transmissions and potentially give their positions away to a listening enemy. The only cable left intact was the German-American line to Liberia and Brazil, and that was cut in 1915. All urgent messages to diplomatic, naval and military stations now had to be sent by powerful radio transmission from Nauen radio station. Soon radio researchers of the British Marconi Company began to pick up an increasing number of radio signals they identified as German naval communications. They immediately brought the text of those messages to senior officers at the Admiralty, who were initially at a loss as to what to do with them.

Transmissions began to be intercepted regularly by hastily erected listening posts on the east coast of England in the first days of the war. Listening, or 'Y', stations soon built up into a wide network of establishments dotted up and down the length of the English, Scottish and, later, Irish coasts. Intercepting German transmissions grew into a major surveillance system and the British also set up 'Y' stations down in the Mediterranean later in the war. The crew of the *Telconia*, and probably the British government, would not have realised that they had begun the work of Britain's code breakers.

Attacks on communications were not just limited to the British as German troops also raided a number of British and French cable stations, cutting overseas connections with India, as well as attacking cable stations on the East African coast which linked Britain's far flung empire. Baltic Sea cables were also severed by the Germans as they came to realise the importance of communications, so exchanges between Britain and her Russian allies became much more difficult.

Cutting Germany's cable communications fundamentally changed the signals intelligence battleground, although neither the Committee of Imperial Defence nor the German High Command could have foreseen its implications. German marine and, to some extent, military communications now became increasingly dependent on wireless transmissions. They were much slower in developing a listening network, but when they had realised that their cable network had been severed they began to develop a wireless telegraphy system to replace it. Nobody had yet realised how vulnerable electronic messages were to interception and decoding, or the part that SIGINT would play in shaping the the First World War.

Wireless Telegraphy Pioneers

Research into wireless telegraphy technology developed seriously in the 1800s, but early progress was mainly centred on the scientific aspects of this new mode of communication. It wasn't until later in the decade, after the scientific principles had been established, that it led to an increasing number of commercial products and services being created. Outstanding among the technology's pioneers, and destined to become a principal figure in the field, was an Italian named Guglielmo Marconi, who had begun to turn workbench-based experiments and pilot projects into working devices. He improved the transmission and reception of his wireless transmission device from a few hundred yards to miles and then hundreds and, finally, thousands of miles. He provided the expertise to build, operate and service the wireless telegraphy products he developed and, just as importantly, the business drive to create a market for them. Marconi rapidly grew from a radio 'nerd' working in his attic on a technology that no one had heard of, into a hard-headed business man negotiating with governments and building the Marconi international brand. He became a dominant force in the wireless telegraphy industry and, although he had many competitors, great and small, none made their mark with the public quite like he did. He and others in the field promoted wireless telegraphy on land and sea, and even in the air, in the first decades of the twentieth century until, by the 1930s, it began to be part of everyday life. The story of early wireless telegraphy was largely the story of Marconi because he brought wireless telegraphy to the market for good or ill. He himself said, 'have I done good in the world or have I added a menace?' The following pages seem to indicate that the industry that he did so much to build has proved a mixed but essential blessing to the military.

Marconi's interest in electro-radiation, or radio waves as they are now known, enabled him to get a place to study the emerging technology and its science at the University of Bologna in Italy. He was inspired by the research of Heinrich Hertz and others in Germany and from this he developed his own early experimental devices enabling him to design a wireless telegraphy system for sending signals a distance without wires. It consisted of a simple spark-producing transmitter, a wire as an aerial, a coherer receiver and a telegraphic key to operate the transmitter that sent the dots and dashes of Morse code, and a telegraphic register that recorded those Morse letters on a paper tape. His initial device could only receive signals over a short distance, but he proved that it worked and that the transmission and interception of Morse code was a feasible proposition.

By 1895 Marconi had lengthened the antennas of his sets, enabling him to transmit his messages for over a mile. He needed money to develop the capability of the technology, so he wrote to the Italian Post Office describing what his device could do – he was promptly invited to take himself off to a lunatic asylum. This rebuff pushed Marconi to take his wireless telegraphy device to London to seek interest and funding for his work as he spoke excellent English. He was able to demonstrate the machine to an amazed audience at the Post Office in London where a plaque on the wall of BT (British Telecoms) Centre still commemorates the first public transmission of wireless signals. This was followed by a number of demonstrations to the British government who invested in the new science. By March 1897, with the increased funding they gave him, he was able to transmit his signals for a distance of nearly 4 miles. This was followed by his first transmission across the English Channel to France in 1899. Later that same year, the American liner SS *St Paul* was the first ship to report her imminent arrival in Southampton using a Marconi wireless set at the Needles on the Isle of Wight while the ship was still almost 70 miles from port. Wireless telegraphy caught the interest of ship owners around the world but, in particular, the world's navies, who saw beyond its commercial use to the way it could be used in war. The naval manoeuvres of 1899 transmitted messages from ship to ship over a distance of almost 100 miles. The Royal Navy immediately placed an order for thirty-two of Marconi's wireless telegraphy (W/T) sets, and the Italian Navy followed suit with an order of twenty – Marconi was on his way. His next step was to span the Atlantic, so in December 1901 he set up his transmitters connected to a 500ft antennae supported by a kite in his attempt to span the ocean with his Morse code message. It consisted of the letter S in Morse transmitted as dot, dot, dot continuously, hoping that it would travel over 2,000 miles from Ireland to Newfoundland. There was some doubt at the time that the signal could be distinguished above the atmospheric noise, but when he announced his scientific advance the public believed him wholeheartedly and Marconi took his place in history.

A couple of years later, President Theodore Roosevelt was able to send greetings to King Edward VII transmitted from Marconi's Glace Bay Radio Station in Nova Scotia, marking the first wireless message transmitted from North America. A dramatic confirmation of the benefits of the new technology came in 1912 when the *Titanic* famously struck an iceberg at speed in the North Atlantic and began to sink. On the stricken liner the two radio operators, Jack Phillips and Harold McBride, who were Marconi International Marine Communications Company employees, sent their distress calls that summoned the RMS *Carpathia* to pick up the survivors of the tragedy. Marconi's Glace Bay Radio Station was the first to receive the *Titanic*'s distress call and it was another of Marconi's men, David Sarnoff, who received and published the names of the survivors from the *Carpathia* via Marconi's wireless set. Wireless telegraphy suddenly became a 'must have' item in vessels of the world's navies and also most large liner and merchant ships that sailed the oceans of the globe. Marconi gave evidence concerning the use of wireless telegraphy in the *Titanic* disaster at the official inquiry in 1912 recommending wireless telegraphy procedures that he felt were needed during emergencies at sea. The British Postmaster General stated afterwards: 'Those that have been saved in the disaster have been saved by one man, Mr. Marconi and his marvelous invention.' It is sure that without the SOS wireless distress signals that the *Carpathia* received from the *Titanic* its sinking would not have been just a disaster but probably also the world's greatest maritime mystery. The great liner would have disappeared among the ice flows of the dark North Atlantic with all on board without trace or sound.

Wireless messages could be received by anyone with the right equipment and the new means of communication was to prove a boon to mankind in many ways. The blessing extended to the military, which needed to communicate confidentially with its men on the ground using coded methods of transmission. Codes and ciphers have made it possible to maintain secrecy in messages and the military were able to send that coded form in Morse to their controllers, whether in London or Moscow. The code breakers were yet to arrive on the scene.

The History of Codes

Encryption of messages into coded form is a science and, like so many other scientific disciplines, has its roots in the culture of ancient Greece. In the Greek language *kryptos* means 'secret' and *graphos* means 'writing'; thus we have the science of cryptography. There are earlier examples of codes among the ancient Egyptian, Hebrew and Asian scholars, but the main tradition of European encryption seems to have come down from the ancient Greeks and was later expanded by the Romans. Coded messages have been used, principally by military men, through the centuries. The first of them recorded their coded messages by stylus and then on papyrus, and in later centuries quill pens and parchment. Within living memory, pen nibs dipped in inkwells inscribed copperplate writing on to paper in laborious hand-written text. Now, of course, it is the universal computer that records and even solves the code using methods built on algorithmic principles devised by the ancients. The science of cryptography 'stands on the shoulders of giants', like so many of our other intellectual and scientific achievements, therefore a bit of cryptographic history is needed to help the reader appreciate how the 'Black Art' of encoding and decipherment of messages has developed and some of the ways it has affected our modern world.

In 405 BC the Greek General Lysander of Sparta was said to have received a coded message called a *scytale* in a servant's belt that enabled him to surprise the Athenian Navy at Aegospotami. The encrypted message was read by winding the belt around a wooden baton which the messenger also carried with him to help the general reveal the hidden text. When Lysander was warned of the Athenian threat he immediately set sail to surprise and defeat them. Alexander the Great used codes and even devised the first postal censorship system while besieging General Memnon of Rhodes, who was fighting for the Persian King Darius at Halicarnassus in ancient Greece. Alexander wanted to know the state

of his troops' morale, so he encouraged his soldiers to write home and then intercepted the courier to read the letters to see how his men truly felt about their army's conditions. By reading their private correspondence, he was able to identify many problems and grievances which he was able to put right, and also to identify the unreliable men in the ranks whom he sent home. Other Greek military men devised an encryption method that substituted letters for numbers using a table called the Polybius Square (below). Using the table shown as an algorithm, the key enables the reader to decode a text in which the letter A is represented as 11, B is 12 and so on. Thus enciphering the word 'war' using the table gives the coding of 52 11 42. A similar code to this based on the Polybius Square was still being used in the trenches of Northern France in the First World War over 2,000 years later.

	1	2	3	4	5
1	A	B	C	D	E
2	F	G	H	I/J	K
3	L	M	N	O	P
4	Q	R	S	T	U
5	V	W	X	Y	Z

In the fifth century BC Herodotus, known as the Father of History, left us an account of a bizarre method of secret writing. It was to shave the head of a slave and tattoo a message on his scalp, and then, after his hair had grown again, the poor slave was sent off through enemy lines to get a hair-cut and so deliver the message. Julius Caesar is said to have used a simple substitution code algorithm like the one described, but the key to his code was to replace the letters in his message, not by one place in the alphabet but by three places. Thus A became D, B became E and so on, so the word 'Caesar' would then be encoded as 'Fdhvdu'. The decrypt would be pretty easily worked out as long as you knew your alphabet backwards. This substitution cipher is one basic form of today's modern encryption system and, as a result, cryptologists refer to this form of coding method as the Caesar Shift Code.

After the fall of Rome, writing as an art decreased and, consequently, so did the need for secret writing and codes until about the fourteenth century when Italian banking families began using simple codes for commercial purposes. More complicated ones were being devised by diplomats in Europe to transmit confidential reports to their masters back home about the ever more political dramas in the courts of Europe. Elizabeth I's court had a most efficient state security service headed by Sir Francis Walsingham. Not only did his service devise and use their own codes and ciphers, but they also broke other people's codes. The queen's spymaster had developed what was for the time an effective coding and decrypting bureau. Plots and counter-plots, mainly by Catholics who wanted to depose or assassinate Elizabeth, were being frequently discovered by Walsingham's agents who intercepted messages and decrypted them if they were in code. An alarming plot was uncovered by Walsingham involving Mary Queen of Scots, who was a Catholic figurehead and at the centre of most of these plots. She had a cipher clerk, Gilbert Gifford, who was arrested at Rye Harbour and 'persuaded' to act as a double agent. He was instructed by Walsingham's agent Thomas Morgan to spy on the Queen of Scots. Gifford was to act as an *agent provocateur* and encourage coded communication between the imprisoned Scottish queen and her Catholic colleagues outside her prison. Most messages were intercepted on their way to the unfortunate queen and painstakingly deciphered by Walsingham's cryptologist, Thomas Phelippes, revealing details of a plot to murder Queen Elizabeth. This early example of code breaking saw Mary executed in 1587 and, later, led to a train of events that caused Philip of Spain to launch the Armada against England.

Daniel Defoe is best known as the author of *Robinson Crusoe*, but is also remembered as the father of the British Secret Service. He was a prisoner in Newgate Prison in 1704 when he wrote to the Lord Treasurer of England, Robert Harley, proposing that he should form a secret service to spy on the gentry of England. The devious Harley liked the idea and got Defoe out of jail, setting him to work to provide a network of spies and agents linked by coded messages throughout England and, particularly, Scotland. At the time the Act of Union between England and Scotland had not been ratified by the Scottish Parliament, so Defoe was asked to convince the Scottish business community that they would grow rich by trading with England if they entered the Union. He personally influenced the Scottish Parliamentary vote to enter the Union with England by pamphlets and promises when it seemed that, in general, Scottish public opinion was against it. He was a major figure in forming the Union as he influenced political opinion to accept the concept which must have enriched many people, although he died unrewarded and in poverty.

In the later years of the Napoleonic Wars, Napoleon developed his 'Great Paris Cipher', a complex code used to pass orders to his marshals, receive reports and direct his armies. The French cipher was cracked by Wellington's code breaker, Major George Scovell, who collected some coded despatches captured from French couriers with the forceful help of Spanish guerrillas. Scovell spent many lonely nights in his tent by candlelight decrypting Napoleon's orders; the intelligence he gathered from his code breaking helped Wellington to secure his many victories in Spain. The original manuscripts over which Scovell toiled into the night can still be seen in the National Archives at Kew. By 1812 most of Napoleon's ciphers had been broken and much of their contents were read regularly by the British Secretary at War in London. Wellington, while fighting the French in Spain, valued the intelligence greatly and was quoted as saying:

> Although we rarely find the truth in public reports of the French government and their officers, I believe we may venture to depend upon the truth of what is written in a cipher.
>
> Lord Wellington, May 1813

George Scovell masterminded the major code breaking achievement of his time almost single handed; he was probably Britain's first serious cryptanalyst, although Britain was about to produce many more. Europe's battlefields spawned a great many secret codes and ciphers, but it was not the only theatre of war that used coded information: for example, George Washington used ciphers to communicate with his generals and spies during America's War of Independence. However, the seismic change in the science of cryptography occurred in Europe with the coming of wireless telegraphy and its coded transmissions at the beginning of the First World War.

Codes and Codewords

In common parlance a code is thought to be a single form of secret signal or writing, but in reality codes and ciphers represent two different forms of encryption. A code assigns a specific meaning to a word, phrase, flag or light, generally calling for some kind of action. For example, a traffic light transmits a coded message to you when the red one orders your vehicle to stop. When the beacons flared from hill top to hill top in July 1588, England knew that the Armada had been sighted and it was time to assemble the troops. Nelson's Royal Navy used a complex set of codes based on flags; every flag was illustrated in a signals flag book with a specific meaning allocated to it. Each ship's captain had such a code book enabling him to read his Admiral's message flying at the flagship's masthead. This code of flag signals had only been perfected for a couple of years before Trafalgar and, if the battle had been fought just a few years earlier, Nelson would not have been able to signal to his fleet 'England expects every man to do his duty' using just a few flags.

The author recalls most vividly the use of secret codes during the Second World War in the BBC's European Services broadcast every evening. Between 7 p.m. and 9 p.m., and after the news, messages to members of the resistance in Nazi-occupied Europe were sent in code. The agents were risking their lives not only by their espionage attempts, but even by just listening to the radio programme while we at home were tuned into the programme in safety. Every evening the programme started with Beethoven's Fifth Symphony as its signature tune. It began dramatically with four single notes DUM DUM DUM DAHH sounding like the Dot Dot Dot Dash of the V for Victory sign in the Morse code. Beethoven's Fifth became Europe's most popular tune for people in Nazi-occupied countries who would hum or whistle it among friends but anyone found listening to BBC programmes could get a visit from the Gestapo; nevertheless thousands did tune in. V signs were daubed on walls all over Europe, so it was as universal as graffiti is nowadays, but to be found defacing a wall with a V sign at that time meant you would be taken away by the Gestapo. The author listened to the messages of hope sent from the BBC telling Europe's people about the war's progress; then at the end of the programme the announcer gave a series of quite meaningless messages that only certain members of the resistance could understand. 'The harvest is good' was one that we remember; it carried instructions to do some deed of destruction or to meet a plane carrying guns and reinforcements to France or the Low Countries. We never found the meanings of those strange codes, of course, but what we did know was that these were coded orders to anonymous men and women to do brave and dangerous deeds. One memorable exception was a code phrase about the 'sobbing of the violins'; we were told after the war that it alerted members of the resistance that the D-Day landings would occur shortly. It was an eerie sensation to be a listener to these strange coded messages directed to resistance fighters engaged in Europe's struggle while we sat at home in comfortable security.

A network of agents needed to keep in touch with their home base and did this by a radio transceiver. Being an operator was the most dangerous job in the resistance, as an agent had to go on air to transmit his message for all to hear and his radio waves advertised his location to the direction finders of the enemy. The radio set could probably fit into a suitcase, but it was always in danger of being checked by an inquisitive sentry or a zealous policeman. Nevertheless, the radio has to go with him or her wherever they went as the operator was always looking for a quiet place to code their message and transmit it. The codes the operators used had to be simple as they could not afford to carry around the code keys or any written material that operators in the London control office had available to them. An agent's method of ciphering his message had to be easy to memorise using a simple and straightforward method. The Double Transformation Code was a cipher used by agents during the war and the key to it was a simple phrase. The code phrase could be as basic as THANKS FOR SOMETHING and would be known to both the agent in the field and their controller at their home base.

To encode a transformed message the letters of the alphabet would be numbered, to keep things simple we have selected 1 to 26 where A equalled 1 through to Z which is number 26. The code phrase has eighteen letters and by giving each letter its alphabetical number it looks like this.

```
T  H  A  N  K  S  F  O  R  S  O  M  E  T  H  I  N  G
20 8  1  14 11 19 6  15 18 19 15 13 5  20 8  9  14 7
```

The message to be sent is RAID ON THE TARGET IS HIGH RISK THREE AGENTS MISSING SO FAR. This would be written out so that each of its letters would be under one of the eighteen letters in the agent's key phrase, as below.

```
T  H  A  N  K  S  F  O  R  S  O  M  E  T  H  I  N  G
R  A  I  D  O  N  T  A  R  G  E  T  I  S  H  I  G  H
R  I  S  K  T  H  R  E  E  A  G  E  N  T  S  M  I  S
S  I  N  G  S  O  F  A  R
20 8  1  14 11 19 6  15 18 19 15 13 5  20 8  9  14 7
```

The code would then be made into a string of meaningless letters by taking each column above from the one numbered 1 (i.e. under A in the key word) to the next numbered column in the sequence, in this case 5 (under E in the code word) to create a string of letters, like so:

```
1    5   6   7   8   8   9   11   13  14   14  15  15  18  19   20  20
ISN  IN  TRF HS  AI  HS  IM  OTS  TE  DKG  GI  AEA EG  RER GA   RRS ST
```

So the new code word becomes:
ISNINTRFHSAIIHSIMOTSTEDKGGIAEAEGRERGARRSST

This new 'word' would be then translated into a string of numbers, by giving each letter in the word its alphabetical number, using the form 1=A to Z=26 as before:

I S N I N T R F H S A I I H S I M O T S T E D K G G I A E A E G R E R G A R
R S S T
9 19 14 9 14 20 18 6 8 19 1 9 9 8 19 9 13 15 20 19 20 5 4 11 7 7   9 1 5 1 5 7 18 5 18 7 1 18 18
19 19 20

So the new code ready to transmit is the bottom string of numbers. Then, to make the decryption more secure the operator took the meaningless list of numbers and laid them out in a similar format to the original message. They are given a second transformation to give some poor code breaker a headache trying to work out what the key phrase was in the first place. We will omit the second transform to keep things simple and avoid your headache getting worse. The meaningless jumble of numbers is then divided into groups of five and tapped out on the Morse key accurately and quickly so that an enemy's direction finders cannot locate the agent's transmitter. It would take an expert at decryption weeks to work out the meaning of the message if they did not know the key phrase.

Back at the control centre the operator had to receive and record the message the agent has sent absolutely accurately for decoding. Any inaccuracy in receiving the coded message could render it meaningless. The message would be decoded simply by reversing the agent's process of ciphering beginning with the key phrase they shared THANKS FOR SOMETHING. The controller knew the code phrase had its eighteen letters representing the number and order of columns of number/letters in the grid. He (or she) is able to replace the long string of number/letters and lay them out in the right order in their columns

If you wish to understand the enciphering procedure better but do not want to get into the complexity of numbers just omit the numbering aspect of coding and work out a transformation in letters. A single transform in letters would have taken the German code breakers no time at all to work out. The above is an example of a fairly simple cipher but the variations can get so complex that they can defeat the cleverest code breakers. If you are a radio agent hiding in a cold windswept barn in the country and listening for the Gestapo to arrive, you probably cannot cope with further complexities than that above. Just to make the understanding of the message even more difficult for the enemy you will probably use code words for an operation or an individual within the message.

Try your hand at working out what the message is here:

The code phrase is: DOG NEEDS DINNER

The numbered code sequence is:
3.8.5.18 11.2.23 18.15.9 4.18.9.5 4.14.7.15 1.19.1.11

Code words were universally used in the Second World War to refer to any major operation or unit, so each major military, naval or air operation had a security name allocated to it (the author's code word was BLOSSOM). Ultra was a code name (although it was too secret to be included in the manual) allocated to signals intelligence coming from Bletchley Park and other sources in the British intercept and decryption system. For security's sake it should not have been possible to guess an operation's nature by reading the code word in a document or transmission if it fell into enemy hands. Why someone broke this unwritten rule and assigned the code word NEPTUNE to the Allied landings in Normandy or SEALION for the German invasion of England we shall never know. Every code word had to be approved and listed so that a copy of *The Authorised Code Word Manual* in use at the end of the war was a substantial book containing thousands of code names with their associated meanings. If a copy of the book had been lost, stolen or strayed into unauthorised hands it would enable the reader to not only understand the significance of the code word or phrase but also a description of the operation it represented.

To summarise, a code is a signal of some kind that has a prearranged significance for both the sender and the receiver, and the meaning is held in memory or a code directory.

Interceptions and Encryption

The battle for supremacy between code makers and code breakers began between the armies of Europe before the First World War, when intercept bureaux were devising more sophisticated ciphers and their code-breaking adversaries tried increasingly devious ways to read the messages. Code making and breaking activities would prove ever more important as the war progressed, so cryptological bureaux in combatant and even neutral countries grew from small *ad hoc* groups of 'boffins' to larger and larger specialist military installations as their workload increased. As the importance of their work became more apparent, they were integrated into the hierarchy of military or naval formations and became a part of those organisations. Before the advent of wireless, the problem for cryptanalysts was that there had not been enough intercepted messages to be able to create a body of coded documents to analyse in order to break their ciphers. (Capturing many French despatches, Spanish guerrillas during the Napoleonic wars helped Wellington's cryptographers considerably.) As wireless messages became the standard method of communicating between units, the reverse problem then arose of how to cope with the great flood of intercepted messages. Senior military officers slowly accustomed themselves to communicating by the use of encoded Morse radio messages until they began to fill the ether. Signallers were having to encipher all messages they had to send out and then decoding the answers, increasing a signaller's workload tremendously. It also increased demands on the radio signaller's skills to not only decipher messages but also observe security precautions in his transmissions. The signals intelligence war had begun.

It started with amateur operators who may not have had anything better to do than listen to transmissions which they began to realise came from the enemy. The interception of these transmissions began to be more organised as the war progressed and listening stations were built to intercept enemy transmissions. The Germans built twenty fixed listening stations around their borders as war approached, of which Lauf, near Nurnberg and built in 1939, was an important one. This Fixed Listening Station (*Feste Nachrichten-Aufkarungsstelle* (FENAST) number 00313) in the Wehrmacht network was where Wilhelm Flicke was Director of Analysis during most of the Second World War, although his cryptographic career had started during the First World War in 1917. Lauf began operating just before war began and was originally meant to cover the military radio traffic from Czechoslovakia but around Christmas 1939 much wider targets were assigned to it in both the Middle East as well as France and the English regions. The station's designation was changed in the 1942 major signals intelligence reorganisation to Laufer Haberloh. A direction-finding site was built in conjunction with the listening post and there were some fifty manned intercept positions, all of which had three radios and two Morse tape strip recorders each working round the clock in six hour shifts. Breaking codes was not just an Allied achievement at Bletchley Park, German code-breaking results increased as the war proceeded to quite high levels, particularly at a tactical level, although unfortunately many archives showing their progress were largely destroyed during the bombing. One code breaker of renown who was a cryptologist well before 1914 and known to Flicke in the *Nachrichtenwesen* (German Signals Service) was Professor Foeppi of Munich University. His team of code breakers were, according to his account, able to crack Royal Navy ciphers regularly during the First World War.

There were few alternatives to wireless transmissions over long distances in either war. Local communications over short distances could use a messenger but national governments and their commanders needed to communicate with their warships at sea and their armies manoeuvring in the field as well as diplomatic posts abroad. The threat of a security breach created by broken codes was a risk that the sender could not assess as he sent orders or reports over the airways. An increasing complexity in codes therefore resulted. Different and sophisticated methods of encoding messages were created by taking both codes and ciphers and mixing them together, a system used in signals transmissions by cryptographers. Simpler methods with a commercial use had been deployed by cable services in peacetime. They charged their civilian customers by wordage so transmission stations provided customers with a cable code book. It listed hundreds of frequently used phrases and each phrase was represented by a four digit number in the code book. By this method the cable company offered customers a more cost effective and swifter service.

Speed of transmission was also important to the military. The prompt delivery of a movement order could be vital so the time to encode and decode a message was critical. It became a trade off as to how easy it was to process a message and how secure and uncrackable the code would be to a listening enemy. Sadly, the Royal Navy erred on the side of simplicity and their messages were relatively easily read by the Imperial German Navy at the beginning of both wars. Codes were favoured by the Royal Navy rather than ciphers as they seemed more secure to them and were more in keeping with their naval tradition. Distributing and protecting code books was a simpler matter in naval vessels than with armies, where units might have to be in constant and sometimes surprising movement and at risk of capture. Navies bound their code books with metal plates in the covers to ensure that, when thrown overboard they would sink out of harm's way rather than be captured with their precious secrets. This did not always work, however, as the Imperial German Navy had managed to obtain a British Merchant Navy code book early in 1915 which could have been partly to blame for the heavy loss of shipping at the hands of U-boat captains. Military security of code books took a different tack. They would be needed to encrypt wireless and field telephone transmissions and would need to be as secure as possible. In a war of movement and manoeuvre, wireless was the main communications method, but in positional trench warfare the telephone took over and was widely used. Code books for use in the trenches were made of lighter and more destructible material that would burn well to stop them from falling into enemy hands.

To capture a code book was a great result for any intercept bureau and is something which did not happen often. More frequently it was the smaller mistakes in transmission which proved to be of value to the code breaker. The French *Deuxième Bureau* were very good at picking up tiny hints from enemy transmissions so that when German radio stations changed their call-signs, as they did frequently for security purposes, the sequence of message numbers on their heading remained unchanged. This could be enough to identify the station to which the French were listening. Habit was the major give-away of a station operator. One German divisional radio station always put its sending time at the end of their messages rather than the beginning, as was the usual case with most operators; another habitually asked the recipient in his transmission 'Can you hear me?' Such details disclosed identities and were a great help to the decoders, enabling them to make guesses as to the source and sometimes even the content of the message.

In the First World War, by and large German intercept services did not perform very well and did not show the disciplined security of the French and to a lesser extent the British in their transmissions. A significant signals advantage occured in February and March 1917 with the intercepting of an increasing number of British telegraphy and telephone conversations. The burgeoning traffic indicated that preparations were taking place for a major Allied offensive in the Somme region and breaking the codes revealed the objectives for the assault. General Ludendorff was told that his troops had not recovered from the Battle of the Somme a year before in 1916 and were in no condition to fight. The general planned an operation codenamed *Alberich* involving a withdrawal to the heavily fortified Siegfried Line. Roads, railways and whole villages were razed to the ground as their troops withdrew. The Allies wasted a powerful artillery bombardment on the empty German fortifications and advanced across a devastated landscape with heavy loss from their enemy's artillery. The Allies soon intercepted enough wireless traffic to become aware of the German intention to retreat but were unable to pursue the enemy so the Allied attack petered out with heavy casualties.

Decryption

As wireless technology developed and streams of coded messages were regularly intercepted, so did attempts at decryption of the code in which most messages were shrouded. Coding and decoding of a text has a scientific basis and mathematical constructs which can reside in the grammar matrix of all languages. Although ciphers have acted as a form of secret communication for millennia it has been difficult to disguise the weak points in its many algorithms, such as repetition of patterns in the coded messages. Cryptanalysis, as the science of decrypting coded texts is termed, has always made the transmission of important information vulnerable but those that devise the many code systems have always seemed to suffer from a great deal of optimism as far as the unbreakability of their own code is concerned. Walls have ears, they say, and so do radio networks, so there will always be the possibility that someone else may not only be listening but may even be understanding what is being communicated in spite of a firm conviction of the sender that the code being used is secure.

The German military were the masters of optimism in this respect and only tried major security changes to their codes a couple of times during either the First or Second World War. Anglo-American TICOM intelligence teams interrogating

German cryptanalysts after 1945 were told that the Abwehr believed Enigma might just have been breakable but they never believed that it would be decoded on such a scale as the huge effort that went into the Bletchley Park project. They never envisaged that the daily output of the Wehrmacht's thousands of Enigma machines or the much more important output of the twenty or so Lorenz cipher machines containing Hitler's thoughts and instructions to his senior generals would be read. Machines that created ciphered messages operated in such a complex manner that the Germans believed it was too difficult to break them. However, there is always a weak link in a machine-generated cipher, the mind bending job was to find it. In Enigma it was that the same letter was never used as a substitute for the same one again in any message. This break in the cipher's pattern was 'the way in' for decryption.

After the war, the TICOM team collected much information about German encryption methods and many kinds of signals equipment from members of the defeated German Wehrmacht who were willing to give it all up to them. The author worked alongside some TICOM members on a Top Secret file created by TICOM dated 12 May 1948 among others. The content is reproduced here:

44/48/TOPSEC/as-14-TICOM

FLICKE: Operating Procedure at an Intercept Station

1. The attached is an Army Security Agency translation of a paper written by Wilhelm Flicke, formerly a member of the LAUF intercept station of the Signal Intelligence Agency of the Supreme Command German Armed Forces. German title: *Der Dienstbetrieb bei einer Hoorchstelle* (Listening Post).
2. This is one of a series of papers written by Flicke giving interesting (and sometimes not so interesting) information on the intercept organisation of German intelligence.

Translated; TAM 25 Copies of which this is copy No. 1
A distribution list follows of 25 designations to see this document.
FLICKE: Operating Procedure at an Intercept Station
The entire field in which monitoring by the intercept services was prescribed was divided by the Control Station of the intercept service into so called
Intercept Assignments H– Gebiste = Horchgebiete or Intercept District.
Generally these Intercept Assignments correspond with national states.
Accordingly there was an Intercept Assignment for the Soviet Union, Poland,

France, Italy, and so on. Great Britain was divided into several Intercept Assignments, corresponding to the geographical divisions of the Empire and the regions of North, Central and South America were divided into three distinct assignments.

Each intercept station was assigned the monitoring of a series of intercept assignments. The most important intercept assignments were regularly monitored by two intercept stations, so for example the Soviet Union was monitored from Koenigsberg and Frankfurt/Oder, Poland from Koenigsberg, Frankfurt and Breslau; France from Stuttgart and Nuenster and so forth.

The intercept assignments which each intercept station was to monitor were established in the so-called Intercept Assignment Plan H- Aufgabenplan. They were broken down according to their importance into assignments of 1st to 4th priority. This order of priority was sometimes absolute, sometimes relative; so for example Central America was a 4th priority intercept assignment but France was assigned to the intercept station at Stuttgart as a 1st priority intercept assignment. Poland was a 1st priority intercept assignment for the Frankfurt/Oder intercept station and a 1st and sometimes 2nd priority assignment for the Koenigsberg intercept station, a 2nd and sometimes a 3rd priority intercept assignment for the Breslau Intercept Station. The weight given an intercept assignment varied according to the political situation. So for example Spain was a 3rd priority intercept assignment but when the war in Morocco broke out it was given 1st priority for the duration of the conflict.

As was seen, the assignment of missions was made purely regional and its principle was followed wherever possible up to the last and was always carried further in the course of time by the changing interests. It happened for example that certain Russian radio stations could be heard better in Stuttgart or Muenster than in Koenigsberg or Frankfurt and French shortwave radio stations were often clearly audible in Breslau or Koenigsberg when they could scarcely be heard in Stuttgart. For these reasons a technical survey of reception conditions was drawn up [in] which the original regional distribution of intercept assignments was disarranged many times. As corollary to this the intercept operator at an intercept station located in the east did not recognise the radio communications originating in the west and had [to be] given additional training in these procedures. The monitoring of the desired intercept assignments became more and more a special science.

The individual intercept receivers of an intercept station were assigned to particular intercept assignments. There were, for example, nine receivers at

the Breslau intercept station in 1931 (later there were considerably more!).
Of these nine receiver positions I and II were for the monitoring of all
international high powered transmitters falling into the intercept assignment
of the Breslau intercept station (this was then Czechoslovakia, Poland, all
the Balkans, Turkey and Hungary). Positions III, IV, V and VI monitored
the army and air force communications of the main intercept assignment
(Czechoslovakia) while the positions VII, VIII and IX monitored the army and
air forces communications of the remaining intercept assignments.

Traffic was passed by the intercept operators to the evaluation group. Here it
was glanced over and sorted by the area specialists or *Landerbearbeitern* and then
went to the individual intercept operators for his information in order that it
might be used for the next monitoring.

Normally the Evaluation Group of an intercept station submitted
four types of report to the Cipher Section of the War Ministry or REM
(*Reichsdreigsministerium*); these were;

1   Signal Operations Reports or *Nachrichtenbetriebsmaldungen* in which
    were noted all elements of identified radio operating technique of
    the radio communications of the land monitored; these included the
    frequency used, call-sign changes, time of communication, method
    of communication, amount of traffic, type of cipher used etc, newly
    identified circuits, items of interest and the like.
2   Radio Reports or *Funkmeldungen* in which, arranged according to
    country the most important messages heard in clear text were listed. These
    concerned messages of a military nature.
3   VN's were called reliable information sources *Vwrlaessliche Nachrichten* in
    which the cipher text and translation of intercepted messages which had
    been read by the cryptanalysts.
4. DF Reports or *Peilmeldungen*. Bearing reports with notes on the results
    obtained through D/F of the radio stations monitored.

In addition the intercept stations provided through monthly so-called
intercept reports and an activity report a review of the operations of the
station for the current month in its intercept activity. The intercept report gave
a general summary for the *Funklage* of the intercept assignment; the activity
report gave a picture of the intercept status and reviewed special occurrences
of the service operations at the intercept station.

In special cases (i.e. monitoring of foreign training exercises manoeuvres etc) the intercept station produced a special report or *Sonderberrichte*.

All these summaries and reports went to the Central Station of the Intercept Service (the Cipher Section) where they were combined with the reports of the other intercept stations so that a survey of the *Funklage* of all foreign countries could be made. The results of monitoring were engaged with the Navy intercept service (and later with that of the Air Force).

In order to maintain close relations between the Control Station of the Intercept Service and the intercept stations conferences were held twice a year in Berlin at which the head of the intercept station and his chief evaluator attended. The individual sections of the Cipher Section presented their reports and made known their directions and wishes for future work.

Direct Teleprinter connections between the Cipher Section and the individual intercept stations for immediate quick exchange in instructions existed.

This very wordy report on the routine of interception stations was by Wilhelm Flicke; he was a cryptanalyst and that needs a special turn of mind, enabling the analyst to think in abstract terms about cipher problems. Bletchley Park found that the most successful decoders were high-flying academics studying the abstraction of higher mathematics, though there were never enough academics with this kind of qualification either in Britain or Germany. In addition to being a thinker in the abstract, a code breaker needed to be incredibly patient as the work was slow and tedious. Tiny hard-won clues could help a cryptanalyst to crack a cipher, but then a signals operator could change the cipher key at any time as a security precaution and a search would have to start all over again. Little wonder that nervous breakdowns frequently occurred among cryptanalysts involved day after day in such close, repetitive and stressful work.

In Germany there was a constant call for cryptanalysts during the Second World War, though the demand was not as great during the First World War. The reason was that in the First World War the nature of code breaking needed the mind of a wordsmith, while in the Second World War it needed a higher mathematician's skills. Early codes were not so intellectually demanding, but from 1939 there always seemed to be insufficient talent available in the shape of cryptologists in the Abwehr as well as a lack of reliable linguists. The British seemed to have a more imaginative approach to recruitment. Advertising for code breakers was not a possibility, not only from a security point of view but also because it would be difficult to identify and evaluate suitable candidates. An indication of how adept a person could be at the abstruse science of cryptanalysis might be how good they were at completing difficult crossword puzzles. The story goes that the British security service devised a high-level crossword competition in *The Times* newspaper. Besides a small cash prize, there was another more unusual prize for the winners and runners up. This was a visit from a couple of mysterious gentlemen who offered the crossword genius the opportunity to work for the government at a job they could not tell them about for very little money. Most of those approached accepted with alacrity – such was the spirit of the time.

The Abwehr were not only short of talented cryptanalysts but they did not use them to best advantage, mainly because their service was not centralised and focused in the way that the British were in Bletchley Park. In addition, German military intelligence interests were much more diffuse, with widespread engagements on several different fronts of the conflict. These had varying military, logistic and language requirements and problems, creating a greater demand for cryptographic teams and linguists. The Abwehr recruited many of their intercept services staff from the Wehrmacht so, as war progressed and the manpower shortage began to bite, many of the experienced analysts were drafted out of their cryptographic units to active duty in the frontline. In addition, there was the ever present threat of the intrusion of the Gestapo nibbling away at Abwehr functions and responsibilities. In retrospect, it is surprising that German signals intercept and intelligence services performed as well as they did.

Frequency Analysis

One method of deciphering text is to calculate how frequently any one letter of the alphabet appears in the text of a coded message. The letter E, for instance, appears most frequently in English and occurs roughly 13 per cent of the time in all letters in an average text, followed by T at 9 per cent then both A and O at 8 per cent and so on. There is also the frequency of two letters coming together such as Q and U which occurs in about 93 per cent of any text. Using these and other statistics it is possible to make some inspired guesses as to the letter occurring in a clear message represented by the code letter at which a puzzled code breaker may be staring. These letter frequencies differ from one language to another of course, so that in the German language the letter E occurs about 14 per cent of the time followed in frequency by R then N then D then T and S. The grammar of a language also carries clues as well, so that in German the verb appears at the end of a sentence which would make the format of a sentence something like 'Germany, the greatest country in Europe is', which makes for another of the cryptanalyst's clues to break a cipher. Differences in letter frequency can also vary slightly between dialects, such as those German speakers in Germany and those in Austria. A coded text in messages could be analysed to calculate the number of times a code letter or image occurs with the use of a detailed knowledge of letter frequency table statistics. If a code letter occurs near to 13 per cent (or 14 per cent in German) of all the letters in the text then a code breaker might assume that code symbol or letter is a substitute for the letter E in the clear text of the message. A number of aspects of the frequency analysis method can occur in a message that could act as a 'crib' for a code breaker to break into the text of the message.

Encryption and decryption come in many forms, of course, and for many different uses from that used in communications at high political level to diplomatic and military codes. These can differ in style and complexity between a commander's communications with his senior officers and that of the trench codes serving soldiers in the heat of action. However, if the operator is too lax or inexperienced to encode his messages properly, then his transmissions are in danger of being read by others.

2

Intelligent Warfare

Intelligence in Signals

In a strictly military sense, intelligence is information used to the disadvantage of the enemy, and it has two functions: intelligence evaluations give guidance (and no more) to recipients to do the right thing on the battlefield; and/or make the enemy do the wrong one. The right thing is mainly concerned with putting your forces where they will be most effective in defence or attack. To get the enemy to do the wrong thing requires a deviousness that will play on and magnify the fears of the enemy so that he is encouraged to make mistakes. At the end of the English Civil War in 1650, Thomas Hobbes wrote, 'force and fraud are in war the two cardinal virtues' – and so it always has been. The difficulty is, to quote Sir Edgar Williams, who was General Montgomery's chief intelligence officer in the Western Desert, 'Military intelligence is always out of date'. He might also have added, 'and always incomplete', as intelligence evaluations only spell out the possible eventualities in a battle situation and not the whole truth, which can only emerge as the fog of war clears and is usually too late to influence events. Military intelligence provides a framework to organise the gathering, collation and dissemination of information from any source of which signals intelligence has been a principal one.

Signals intelligence is an important management tool, acting to help clarify the intentions of the enemy, and has become a specialised art form contributing to what military intelligence has become in the twenty-first century. The discipline had two main aspects during the First World War, the first being Communications Intelligence, or COMINT, of which signals intelligence is a major component, with field telephones also included as a messaging system. Electronic Intelligence, or ELINT, was soon added, with the main technique being the use of direction-finding, or D/F, equipment to locate the position of enemy emitters. (In more recent times, the jamming of missile control systems by electronic interference can be included in ELINT.) It served as a complimentary tool to COMINT as it was able to fix a distant enemy transmitter's geographical location and help to map the positions of an enemy formation on the battlefield. It also had an increasingly important role in spotting ship or submarine positions at sea as they were sending their wireless messages. Morse code was almost the universal means of transmitting messages in both world wars and is still used to some extent today by military forces of the former Soviet Union. Military manuals describe signals intelligence with a commendable economy of words thus:

- SIGINT is a category of intelligence comprising of, either individually or in combination, all communications intelligence (COMINT), electronic intelligence (ELINT) and foreign instrumentation signals intelligence, however transmitted.
- SIGINT Intelligence is derived from communications taken from electronic and foreign instrumentation signals other than the intended recipients.

Wireless communication using Morse code and, later on, audio broadcasting by radio or telephone, enabled commanders on land, sea and even in the air to direct men, machines, materials and artillery over great distances in battle. Early on in the First World War, wireless was a novel means of communication that created fundamental changes in the dimensions and nature of battle; the technology required confidentiality not possible in clear text messages, so the increasingly dark side of encoding and decoding messages in ciphers and codes began to appear. Intercepting wireless messages and decoding them to try and understand the enemy's meanings and intentions became a major objective for every signals intelligence officer. Being able to identify the location of a transmission by D/F, as well as reading its contents, was almost like sitting with the enemy and listening to their plans. An early technique, which is still used for locating an emitter station, is by using D/F antennae as measuring devices, picking up enemy transmissions and then using compass bearings to plot them on a map from several different D/F stations. This was not as simple as 100 years ago, when they did not carry magnetic deviations or details of the terrain as they do now. Even so, plotting the compass bearing lines from several D/F stations would show where the plot lines crossed to indicate the approximate location of a transmitting wireless set. An approximate position in D/F calculations could mean an error of 100 yards in locating an enemy operator's set 1 or 2 miles behind the lines in France in 1914–18, although, if used at sea, the location of the vessel could be out by 10 miles in the North Sea and 30 miles if the boat was far out in the Atlantic. D/F techniques were not very accurate in the early stages of their development, but the results were better in the hands of a skilled operator using codes and ciphers. An operator's skills in encoding and decoding messages improved as the war went on and so the enhancing of codes and using techniques to confuse direction-finders improved. By the dawn of the electronic age, the battle of wits in signals intelligence war began.

An expanding pool of radio amateurs developed in the inter-war period in what could sometimes become an all-consuming hobby, and when the need began to arise, it was possible to dip into that pool to recruit radio agents and operators. The attitude to amateur radio operators and their operations differed from one country to another in a way that suited their political and even cultural needs and objectives. In Germany, during the Weimar Republic (1919–33), amateur radio traffic was prohibited for fear of communication between the rising Communist Party in their country and the Soviet Union. Hitler's Third Reich continued this policy, fearing increased espionage activity and communication between various subversive elements inside Germany. The emerging need of the Reichswehr (the Reich Defence) to develop the skills of cryptographers like Wilhelm Flicke and his colleagues by practising under supervision was recognised and regulations became more relaxed.

Radio amateurs in Britain were restricted as well, but not to the same extent as the Germans, and the British Radio Amateur Association was used by the government to act as a regulatory body. They checked on the loyalty of its members and tried to prevent abuses such as communicating with undesirables of many kinds, mainly foreigners. Government policy did, however, encourage the members of the association to communicate with other radio hams (amateur radio operators) in the colonies and countries throughout the empire.

The United States had the culture and, indeed, constitutional right of free speech and was less concerned with the risks of subversion among its amateur radio community, so there the radio ham community was bigger and more vibrant. Amateur radio strongly took on a sporting aspect there and provided a market for entrepreneurs to design and produce radio equipment of an innovative nature. This growing body of experienced participants and equipment became available as war was declared, but their enthusiasm did not always make for security of transmission in combat conditions. It also enabled German, Italian or even Japanese immigrants to become radio agents in the USA. Germany was able to recruit many of them in North and South America to become agents for 'the old country'. This became a problem for the Allies before the war as information about British shipping movements was often transmitted to the Kriegsmarine (Hitler's German Navy) from the United States and, particularly, Canada. As Germany declared war on America in 1941, this problem lessened as immigrants had to choose between the interests of their old country and the one that they had adopted.

With the Communist Party in firm control, the Soviet Union's attitude to amateur radio users was predictably different. The policy between the wars was to encourage radio amateurs among the 'reliable' to practise their skills in radio communications, which paid dividends when the war started. The skills in the use of shortwave radios by Russian partisans, whose activities cost the invading German Army so much, had been learned by party members before the war. Likewise, many hundreds of Russian agents who caused mayhem behind German lines served their apprenticeship in radio communications well before the war.

The main purpose of signals intelligence when war came was to gather enough information to support a commander in the field and estimate the Order of Battle, or ORBAT, of the enemy. The ORBAT is the identification of the strength, structure and disposition of an enemy force, including the command structure of that force. The prime task of any intelligence unit within a formation of troops is to identify all of this. Such intelligence units are generally to be found within the staff serving the commander. Any general wants to know the strength of his opponents so his order to his intelligence officers would be to obtain as much detail of the enemy's ORBAT as they can. An estimate of an opposing force's strength, structure, dispositions and command chain would be needed, so he would expect his intelligence section to use all the methods at their disposal to get it. The size and strength of an opposing formation is invariably estimated by painstakingly matching many small pieces of detailed information to create a clear – and hopefully accurate – evaluation of the enemy's situation. An intelligence appraisal of the enemy's strengths and weaknesses will probably act as the commander's basis for judgments on how to make his dispositions and movements.

Initial unsteady steps in the development of telegraphy in both commercial and military usage had already begun at the turn of the twentieth century; some belligerents already had their own signals intelligence experience or even, as war started, a working signals intelligence bureau. Wireless messages in encoded Morse code signals were being transmitted and sometimes intercepted by a growing band of largely inexperienced wireless operators. Skills and lessons had to be learned, sometimes at great cost, particularly as wireless telegraphy became more accepted as, in some cases grudgingly, a normal means of communication. During the First World War, the German Army were very slow in developing both their interception services and practices to keep their transmissions secure. This was in spite of their great success in intercepting most Russian orders and instructions to their troops during the Battle of Tannenberg. This massive advantage should have taught the German commander a lesson; however, Field Marshal Hindenburg did not learn the advantages and, just as importantly, disadvantages of signals intelligence. He was afraid that his army on the Western Front would make similar fatal mistakes to those that the Russians had made. They did fall into the same trap, however, as the French had developed the expertise to decipher and read the signals of the German Army on the Western front. They advanced into France in a similar way that Hindenburg had done in the east, with the French reading their signals all the way. Meanwhile, the Imperial German Navy were losing their code books at a surprising rate, giving the British Admiralty an early and lasting advantage in code breaking. The Royal Navy did not make full use of this advantage, at least in the early part of the war, as the distribution and use of their intelligence evaluations was not what it should have been. On the other hand, the French utilised their decrypts well and handled their distribution during the Battle of the Marne to make sure that the intelligence evaluations went to their field commanders in a timely way.

As the war progressed and trench warfare became more static during 1915, messages were increasingly given over field telephone networks. However they were sent, they all needed transmission security to protect their content and keep them secure. D/F was mainly a British invention and a German system did not appear until late 1915, although the French cottoned on to the principle quickly and began developing techniques of their own. A transmitter's location was important in intelligence terms as they were invariably attached to the staff headquarters of an enemy formation. This could not only give an insight into the enemy's ORBAT, but also an indication of the ground covered by a formation through the location of its operational transmitters and its frontage. If a D/F operator was able to identify more than one of the enemy's emitters, it could also help to trace the movements of the unit, sometimes over quite long distances. Longwave sets were still being used at the time, so it was possible to get a fairly accurate fix on a wireless transmission within a few hundred yards or so of its position. Direction-finding plots rendered the disguise of radio transmission call-signs much less important as a wireless operator found it increasingly difficult to mask his identity if his location was plotted. Tracing army units was particularly useful on the Eastern Front as the distances in that region were much greater than in France and the units more mobile. Tracking Russian transmission stations during their preparations for the Brusilov Offensive into the Ukraine in 1916 by the Germans was a considerable factor in countering that offensive. As a result of the success of D/F, it was used to monitor Romania's attack on Austria and, subsequently, on Austria's grand offensive on the Italian Front in 1917 to good effect.

Many archive copies of First World War German military strategic papers were destroyed by bombing in the Second World War, but reading what was left of intelligence evaluations of the time seems to indicate that they probably suffered as much from the lack of trained intelligence personnel as the other belligerents. German usage and practice of intelligence evaluations later in the war did improve, but their expertise never quite caught up with the French or British intelligence achievements. The war's first few months on the Western Front were just the kind of fast-moving actions that got the best (and worst) out of signals intercepts and intelligence evaluations. SIGINT played a significant part in shaping the war of movement in France, but later both sides slowed to a virtual stop to dig trenches for fighting a positional war. Thus the new environment needed new weapons with different tactics to suit the changing circumstances and interception methods. German listening stations generally had a dual listening and receiving role, and one of the principal ones was a powerful transmitter in Nauen near Berlin, though it did not have the position or enough power to reach far out into the Atlantic. When the Germans occupied Brussels in 1914 they immediately took over that city's powerful wireless transmitter station and made good use of it during the whole of the war, but mainly in a transmitting mode. It would become the scene of the most dramatic signals intelligence leak in the history of signals intelligence, and the war's most important diplomatic event.

Meanwhile, the German Imperial Navy had established its *Entzifferungsdienst* (*E-Dienst*) listening service as an intercept station, receiving messages from both Germany's U-boats and their surface vessels. The station was situated on the coast in the little town of Neumunster, near Hamburg, where the *E-Dienst* transmissions outperformed German Army signals intelligence efforts easily. It did not guess, however, that its codes were being read by the British Admiralty in Room 40. *E Dienst* did have some successes, such as the breaking of a Royal Navy code at Jutland, but transmission and reception principles only established themselves slowly as they gathered experience. In all things maritime and military, a regime of discipline in the use of wireless transmissions and safekeeping of code books would give the best results. Unfortunately, both were lacking for *E-Dienst* and the Imperial German Navy, and every vessel's movement was known in London almost as soon as it prepared to leave harbour. This led to several sea battles but, most importantly, led to knowledge of the position of the U-boats in Germany's fleet at sea. This enabled the Royal Navy to counter submarine movements by redirecting convoys using methods developed in the First World War and used again in the Battle of the Atlantic in the Second World War.

German transmissions intercepted by the British Admirality's listening 'Y' stations came mainly from the Kriegsmarine's high-powered transmitter in the tiny North Sea coast town of Norddeich. Other less powerful wireless transmission stations in Kiel were also listened to with close attention by the Admiralty. Indications of activity concerning the movements and putting to sea of the Imperial German Navy, no matter how small, were of prime interest to the Admiralty and its Grand Fleet berthed at Scapa Flow. Identifying the wireless stations of interest to the Admiralty was the work of 'Y' stations on twenty-four-hour watch on all known wavelengths the Germans used. Teams of operators sat in rows to listen to an array of co-ordinated receivers tuned into specific wavelengths allocated to each operator that would be on continuous alert for the first burst of Morse code crackling in their ears as an enemy operator began their transmission. The location of the transmitter was as important at sea as on land, so D/F units were also on alert to locate the position of a ship or submarine at sea or shore station sending fleet instructions. The British Admiralty in London used Room 40 in its building to co-ordinate the signals intercepts while the outstations monitored the German radio traffic and any movements of blockade runners. Operators on both sides devised signalling tricks to try to fool their distant listeners with frequent changes of wavelength and other countermeasures used against interception, as well as other tricks to confound unwelcome listeners. Many kinds of safeguards were available to operators even before they started to encode the message. *E-Dienst* continued to operate right up to the end of the war and even reported the mutiny of the sailors of the Imperial German Navy in 1918 which did so much to bring the war to an end. Ultimately, the Imperial German Navy operators did not compare with the skill and organisation of the occupants of Room 40 in the signals battle at sea.

The inter-war years revealed to the world through memoirs, lectures and books the tremendous advantage that Britain had gained in the use of signals intelligence. As a result, Germany's Abwehr, or military intelligence section, began to build a signals intelligence capability, as did the German Naval Staff, or *Admiralstab*, who set up its *Beobachtungsdienst* (*B-Dienst*) intercept service. The Japanese took lessons in cryptography from the Royal Navy until 1937 and Russia improved her intercept services out of all recognition. The Spanish Civil War (1936–39) was a dress rehearsal for the war that was yet to come, and a German intercept company gained much experience while fighting for General Franco and proved the effectiveness of their equipment. By the Second World War the whole range of technology had naturally become much more sophisticated and complex, although Morse code was still the standard method of contact used by the military and, in particular, the navies of the world. Wireless telegraphy, or W/T as the Royal Navy called it, was the armed forces' chief means of communication for all nations. Signals intelligence added another dimension when the new ELINT techniques, such as radar and voice transmission, began to replace the dots and dashes of Morse transmission. The future of ciphering messages took a leap forward as encoding and decoding became more mechanised. Examples are the Japanese Purple, the American SIGABA/ECM and the German Lorenz and Enigma machines.

The growing size and sophistication of electronics warfare required a methodical way of handling raw data from the intercepts and turning it into intelligence evaluations that were of use to a battlefield commander. The commander needed his intercept services to produce an evaluation of the enemy's strength and capability in an ordered and reliable manner. Establishing an opponent's Electronic Order of Battle, or EOB, as a part of his overall order of battle required an intelligence unit to identify the number of enemy wireless emitters in the geographical target area of the enemy formation. Using D/F techniques it was possible to establish the radio stations, both mobile and static, and endeavour to identify the characteristic procedures of each enemy radio station and the unit's role in the chain of command. An example of how a command structure might be traced is to imagine that a transmission from an enemy station had been intercepted and the messages decoded. The message might request permission for the unit to move to a fresh location or some other submission. If the second unit approves the move then a link to the senior unit has been established. If the senior unit then reports the action to a third unit, a chain of command has been established for the three stations. A part of the British Eighth Army's higher command chain was established during the desert campaign by intercepting requests to dispose of certain documents out of filing cabinets. To gather such transmissions, huge aerials were used in the listening posts, backed up by smaller ones. These could be focused on distant and sometimes indistinct enemy radio transmissions as a first step in forming such an intelligence picture. Radio stations are invariably attached to a specific army formation so the listener may identify the name of the unit, while D/F scanning would try to establish its location which, in turn, might help to identify its military function.

In the Second World War, the *B-Dienst* signals intelligence arm of the Kriegsmarine was able to use wireless and listening stations on the French coast, and these proved to be the principal adversary of Bletchley Park. The Wehrmacht did not have a centralised cryptographic and intelligence centre as Britain did in the 'Park', but the Kriegsmarine, the Luftwaffe and the *Heer* (the German Army) each had their own intelligence organisation. The German Navy's intercept service performed well in both wars and their operator's training syllabus enabled them to recognise the Morse key style of individual telegraphists in important stations or ships. The technique was called 'finger printing' and could sometimes enable a German operator to identify some of the Royal Navy's ships or shore stations operators.

The Luftwaffe badly underestimated the order of battle of the Royal Air Force (RAF) Fighter Command in 1940, including its 'Home Chain' command structure. This was an information network that was feeding data into a central station to control the fighter squadrons at their airfields. The structure was based mainly on the new ELINT technology of radar, but with several other aspects such as the Observer Corps adding human intelligence to the network. Hermann Goering, as head of the Luftwaffe, might have fought the battle differently if he had a better intelligence-led appreciation of Britain's air defences. The nature of Britain's Home Chain defences and information system was a crucial part of Britain's defences and yet was overlooked by him. A misleading intelligence evaluation of Fighter Command's order of battle made by the Luftwaffe before the battle assessed the RAF strengths in its control network, and the evaluator apparently did not highlight any weaknesses. The weakness of the RAF's rigid air discipline in fighter formations, however, was evaluated as a strength. As battle commenced, there were many clues as to the effectiveness of the 'Home Chain' system; Luftwaffe wireless operators listened in awe to the calm voices of RAF controllers as they 'got up' their fighter squadrons just in time to intercept the bombers. It must have dawned on the listeners that they were witnessing the product of a very efficient information network, able to vector fighter interceptions on to most of the Luftwaffe's raids. The same was true of German blockade runners in the First World War, which were often intercepted by British warships. German intelligence evaluators must have been puzzled by the fact that the Royal Navy was so often in the right place to intercept German blockade-running merchantmen. They did not seem to suspect their wireless security could be compromised by a leaky signals coding system, to their great cost in both wars.

The new generation of signals intelligence that evolved during the Second World War had made the technology hugely complicated as radar was added to the ELINT technology and voice transmission to COMINT. In spite of this, even at the end of the Second World War the transmission of messages was still dominated by the use of Morse. Its universal use created a huge body of skill among operators who could pick up all sorts of hints and clues from enemy transmissions, particularly if they monitored them regularly and became familiar with their pattern of sending. One aspect of this scrutiny was traffic analysis, a signals technique that identified an enemy's preparation for action merely by measuring an increase in wireless traffic. One aspect of traffic analysis was employed by the German Abwehr to assess the intensity and frequency of transmissions created by secret agents operating in Occupied Europe. They were able to recognise the areas where the Allies were planning new operations. In May 1944, German intercept stations noticed an increase in English agents transmitting from the islands of the Aegean and, in October, the islands were occupied by the Allies. In June 1944, a centre of agent activity sprang up in the south of France and, a month later, Operation Anvil landed a major Allied invasion force on the Côte d'Azur. The main incursion of agents occurred, however, on the Russian Front when agents were parachuted into Hungary; increased transmissions indicated that the Red Army would be attacking in that region before their last main assault on Berlin. The prize for using signals traffic management went to the Allies as D-Day approached; their remarkable control of radio transmissions gave the German defenders on the coast of Normandy no clue that a huge Anglo-American invasion fleet was on its way.

Both sides searched the airways and the wavelengths to pick up faint Morse code or sometimes voice transmissions from distant radio sets. German Luftwaffe radio operators in France listened to conversations between Spitfires over Kent and their RAF controllers directing them during the Battle of Britain and got some useful knowledge from it. German listening stations seem to have been manned by men that had been wounded or those whose age made them unfit for more active duties, but it did not seem to affect their ability to listen. Nor did the Abwehr employ women to any great extent on intercept duties until late in the war. Indeed, women did not seem to be recruited as widely into the German armed forces as they were into the ranks of the Allied forces. Both German and British listening posts seem to have been less than effective. It has been said that British listening posts improved when American sets were imported, while other sources claim that receivers from British Telecom were of high quality. German sources remarked that British listening posts could pick up signals from South Africa but reception was unreliable. There seems to have been less co-operation linking German listening stations than in Britain, which was important in matters such as direction-finding etc. Even so, the Germans were able to identify and locate every military unit on England's South Coast in the build-up to D-Day, just by listening to military police using their personal radios to direct military traffic. Similarly, television transmission, which was closed down for public use in Britain at the beginning of the war, was still being used for military purposes, and these black and white transmissions were being received in 1944 by the Germans on the other side of the Channel. Allied soldiers in England did not realise that Abwehr signals operators could be watching their every move. At the same time the RAF were monitoring German TV broadcasts in France from Beachy Head in Sussex.

Signals intelligence was slow to be accepted by both politicians and commanders on land and at sea in the early stages of both world wars, but this improved considerably as the war progressed. An evaluation of the significant raw data and volume of transmissions in the First World War, and, to a lesser extent the second, took several months to indicate to commanders that something was up. It took the disaster of the sinking of HMS *Glorious* in 1940 to focus the minds of senior naval officers on traffic analysis evaluations. Its value was known to cryptographers and when a certain wavelength became busy it made them suspect that the German heavy cruiser *Scharnhorst* would be putting to sea. The message was discounted by officers at the Admiralty until the German warship engaged and sank the British aircraft carrier HMS *Glorious* with all hands while returning from Norway. After this incident, naval officers treated all evaluations by Bletchley Park cryptographers with great respect and high priority. No such powerful lesson was given to the German General Staff, so they had to learn the value of signals intelligence more slowly.

Many on the German General Staff were dismissive of signals intelligence, and indeed all intelligence evaluations to some extent, because of the struggle for dominance between the Abwehr and the SS interception service: the *Sicherheitsdienst* (SD); a parallel SIGINT service created between it and the Abwehr meant that both could be intercepting the same message, both decrypting that same message and both making an evaluation which might give different values to an intelligence source. This did not engender confidence. Differing evaluations sometimes went to the same people, so there is little wonder that German senior officers found it difficult to accept the advice or guidance from such sources. In spite of this, the Abwehr intelligence service maintained a high standard until almost the end of the war. The naval signals intelligence service did not have the same problems as their war at sea was of no interest to the Gestapo. The Kriegsmarine therefore had no competitor and were able to chalk up some real operational intelligence successes against ships of the Royal Navy, British Merchant Navy and, later, the US Navy.

Senior officers on the Allied side used intelligence appreciations well in the second part of the war as signals intelligence got into its stride in 1942. Evaluation reports helped in some other successful actions. German generals treated intelligence from central authorities with reserve, as we have seen, but they used their own tactical listening units to gather mainly local military intelligence; Rommel utilised this method to good effect in North Africa. You might think that generals would be grateful to their subordinates for the good intelligence which made their successes possible, or at least easier, but that was not usually the case. Generals usually had a significant ego, otherwise they would not have made it to the military rank to which they had risen. In the many accounts we have of actions in which signals intelligence contributed to the successful direction of a battle, there is rarely more than just a mention of how helpful a timely message might have been to the man in command. One outstanding example of this was von Hindenburg's victory at the Battle of Tannenberg and the contribution made by signals intelligence. After the war he wrote his memoirs, entitled *Aus Mienem Leben*, in which the many intercepts he is known to have had were ignored and did not help him in his immense victory.

Politicians and Intelligence

It is clear that SIGINT was often undervalued and misunderstood. To really excel as a weapon of war it needed a champion. In the case of Bletchley Park, its outstanding champion was Winston Churchill. He had learned the value and use of signals intelligence when he was First Lord of the Admiralty in 1914, resulting in the establishment of the Admiralty's cryptographic bureau, Room 40, which was to be the forerunner of Bletchley Park. Churchill's depth of practical and even emotional experience is shown in his personal account of the scene as radio messages are received at the Admiralty. The British High Seas Fleet were about to meet a force of German warships scouting the Dogger Bank as dawn broke over the North Sea, an encounter assisted by signals intelligence. The story is told as only Churchill can tell it:

> There can be few purely mental experiences more charged with cold excitement than to follow, almost minute by minute, the phases of a great naval action from the silent rooms of the Admiralty. Out on blue water in the fighting ships amid stunning detonations of the cannonade, fractions of the event unfold themselves to the corporeal eye. There is the sense of action at its highest; there is the wraith of battle; there is the intense, self-effacing physical or mental toil. But in Whitehall only the clock ticks and quiet men enter with quick steps laying slips of pencilled paper before other men equally silent who

draw lines and scribble calculations and point with the finger or make brief
subdued comments. Telegram succeeds telegram at a few minutes interval as
they are picked up and decoded, often in the wrong sequence, frequently of
dubious import and out of these, a picture always flickering and changing
rises in the mind and imagination strikes out and around it at every stage
flashes hope or fear.

Winston S. Churchill in his memoirs of the First World War

His vivid description could just have easily related to the unfolding action at the
Battle of Jutland later in 1916, the pursuit of the *Bismarck* in 1941, or the disaster
of Convoy PQ 17 in 1942; all of which were played out miles away from Britain's
shores, but relayed in intimate detail through a series of signals and codes.

It was Churchill's habit to telephone the duty officer at the Park in the
small hours to ask how his hens were laying their eggs (he sometimes added
that they should not cackle), meaning of course signals intelligence from the
decryption Enigma and, particularly, Lorenz coding machines. Churchill's solid
commitment to the benefits of signals intelligence led to the huge investment
of effort and funds in Britain's intelligence network, of which Bletchley Park
was only a part. The Park's achievements were not only cryptographic but
also the way that intelligence evaluations were handled and distributed to
commanders in the field. Some generals had to be encouraged and even bullied
into using Ultra intelligence evaluations, as some 'old school' officers never saw the
value of them. It was Churchill's driving force behind Bletchley Park that made
Britain's cryptographic achievements and intelligence gathering so effective.

Hitler, on the other hand, had little experience of military intelligence and
almost none of signals intelligence, so he did not greatly value its contribution
to his plans. Indeed, he spurned the unwelcome evaluations with which he was
presented towards the end of the war.

Generally, though, there seems to be little 'feeling' for the almost instinctive
practice of active intelligence in German culture; no word in the German language
easily translates into English to carry the meaning of intelligence in the military
sense. During the Second World War, the intelligence arm of the *Oberkommando
der Wehrmacht* (OKW) was the Abwehr. The term translates roughly into the
word 'defence' in English and the Abwehr's brief did not have an overall concept
and intuitive understanding of information that so successfully drove the British
intelligence service. All early security agencies in both Germany and Austria in
the First World War started their lives focused on the need for home security, and
this mindset continued into the Second World War. They always had the tendency
to try to control their own people, which is a problem that British intelligence

did not have. The Head of Operations of the Abwehr, Admiral Canaris, was an exception to this trend and had an understanding of the uses of intelligence because he had practiced the art in the First World War. However, his political masters did not understand and few of the Admiral's subordinates in the Abwehr understood either. Canaris recounted how he described his organisation and its purpose to Hitler in 1935, soon after the Nazi Party came to power. Hitler said that the British and Soviet intelligence services were far superior to Germany's but made no suggestions as to how to improve it. Asked about the American intelligence agencies, Hitler dismissed them as there was only the FBI and they had little interest in international matters. He was right about that. The Führer expressed more interest in international gossip such as the affairs of the British Prince of Wales with married women, the relationship of Reynaud in France with his mistress, and who was bribing who among the various politicians of Europe.

Hitler did intervene disastrously in the Abwehr administration later in the war by ordering the limitation of all intelligence evaluations to a short list of his very senior administrators. Often twenty pages a day of valuable Abwehr intelligence was generated and available in a day for circulation to senior officials on the list. As busy people, they were frequently unable to do more than glance at the detailed contents of this unceasing flow of paperwork. Moreover, they were forbidden, on Hitler's order, to share or pass on the contents to others, so many overworked officials could not delegate the study of intelligence evaluations to any of their subordinates. As a result, German intelligence and intercept services reports were often initialled and passed on to be filed away. At the same time, German civil and military administration were required to destroy all confidential documentation every three months, so many of the Abwehr intelligence evaluations were incinerated. The Führer's system therefore ensured that the most senior administrative staff running the German war machine were not studying their intelligence briefings.

Hitler's lack of interest in the operations of his many intelligence services meant they lacked direction. He allowed seven intelligence agencies to spring up in the inter-war years and to stay separate and fight it out among themselves for dominance and power in a way that contributed to a confusion of intelligence evaluations.

The exception was the Kriegsmarine, which ran the most effective intelligence operation and, as a result, was unusually suspicious of the security of their codes. The Wehrmacht, the Luftwaffe, the Gestapo's SD and all the other various intelligence agencies stuck rigidly to the belief that their communications systems were impregnable. They were watching each other and not the enemy, so their eye was never properly on the ball. The two main protagonists in this inter-rivalry were the German High Command's Abwehr intelligence arm and its counterpart agency and foe, Heinrich Himmler's SD cryptographic service, or the dreaded SS. The SD grew to be so powerful that it finally took over the Wehrmacht's intelligence arm in the struggle for power within Germany, but did much damage in the interim. Both organisations were allowed to have responsibility for forms of internal security, but later in the war the SS expanded its power base into international affairs. Competition between the Abwehr and the SD duplicated the efforts of the two intelligence organisations, but even worse was the time wasted by the various agencies plotting the downfall of the others. Inefficiency and time wasting by this unproductive state of affairs caused the encryption services of the Third Reich to carry a crippling burden, as not only were the various intercept services of the Axis powers at each other's throats, they also had a greater spread of enemies to combat abroad. British and American forces were poised to assault German defences on the French coast from the west, the Red Army was advancing inexorably towards Berlin in the east, and the desert war and the Balkans in the south all posed a growing range of problems for the German intelligence services. As if that was not enough, there was the major irritant of the many subversive activities of foreign secret agents and their radio transmitters operating within Occupied Europe. For example, the problems caused by Russian partisans operating behind German lines in occupied Russia, controlled from Moscow by shortwave radio, ultimately contributed to the collapse of a whole Army Group Centre on the Eastern Front in 1944.

3

The Pre-War Intelligence Scene

Russian radio communications were a disaster in the First World War, but had improved greatly in the second; the disastrous misuse of transmissions early in the First World War certainly altered the shape of the war that followed. The Abwehr recognised that Russian radio transmissions during the attack on Poland had high standards of security, although the surprise German offensive against Russia in 1941 cost the Red Army many of their best operators, who they found difficult to replace immediately. However, the standard of radio performance was well maintained, at least in the higher echelons, which may have been due to the fact that Marshal Stalin had observed the importance of radio procedure while serving as a Commissar in a Communist cavalry unit during the 1920s. Stalin gave impetus to the army's improvements in radio techniques and security, so that radio discipline was strictly enforced on all ranks by the NKVD (Soviet police) and infringements were punished, often severely.

German Intelligence

A German intercept and decryption service started producing results quickly and achieved some notable successes early in the First World War. The most successful operators in both world wars, however, were the German Navy, whose U-boat fleet came near to strangling Britain's seaborne supply routes.

The German Army's development of its intelligence function began in the First World War under the direction of Colonel Walter Nicolai. He headed his department from 1913 to 1919, but did not achieve much in the way of signals intelligence evaluations in the first year or so of the war. The German High Command had little access to a workable signals intercept service to speak of until the end of 1915, and military intelligence never properly caught up with their opponents in this field until near the war's end. Nicolai was a professional army officer with a Prussian background and attitude, and was appointed to be the first senior intelligence officer in the German Army. He led the Military Intelligence Section, known as Abteilung IIIb, throughout the war and seems to have achieved some success. Espionage was, however, a different game at that time. Some of his agents achieved notoriety; for example, the famous Dutch exotic dancer Mata Hari and the Irishman Sir Roger Casement, both of whom were caught and shot as spies. Nicolai's activities were mainly directed against France, although his bureau also targeted Russia, so it helped that the colonel was fluent in both languages. Nicolai was fiercely nationalist, as most senior intelligence men seemed to have been, and played his part in passing on to others some of the skills and tradition of signals intelligence that he had developed. At the end of the Second World War the Russians believed that the colonel was a part of the Nazi intelligence system and arrested him as they advanced into Germany in 1945. He was taken to Moscow for interrogation by the Russian KGB (secret police) and died there. He was always a hero of Admiral Canaris of the Abwehr who had his picture hanging in his office.

Colonel Nicolai, as the first senior intelligence officer in the German Army, was posted to Section IIIb in the Koenigsberg Fortress in East Prussia in 1906. There he created the German Army's first intelligence centre for espionage, mainly spying against the Russians. By 1913 he was head of German Army intelligence. It was his action while commandant of the Koenigsberg Fortress in bringing wireless intercepts of the Imperial Russian Army's clear text messages to Hindenburg's attention that assisted so materially in the German victory at Tannenberg. His exploits in the war seem to be largely unrecorded; the book that he wrote of his experiences is currently out of print. His reputation, according to his contemporaries, was as a dedicated and ruthless exponent of his country's intelligence needs. Colonel Nicolai developed the basic structure of the German Army signals intelligence system, based on separate units of company strength that were able to monitor either a long or short-range intercept service for the divisional formations to which they were attached. Fixed intercept units such as Koenigsberg Fortress would report to the German General Staff and would probably be long range in nature. The structure of signals intercept units that Nicolai devised in the First World War continued into the Second World War.

In reality, signals intelligence in the German Army did not exist when war broke out in 1914, but when it did develop by 1916 it became a structural part of the military intelligence department. Intercept companies, introduced in 1917, were the basic unit, equipped and trained to listen to enemy wireless transmissions. Military formations of army corps size generally had one attached to it, reporting its findings direct to the commanding officer. There was, though, little in the way of centralised interception and almost no conception of organised cryptanalysis (basically, decoding) until much later in the conflict. The opening battles in the first months of the war used signals interception almost accidentally and found that it helped to direct the movements of armies and boundaries of the battle. Wireless messages proved to be less important to armies as they lost momentum in their movements and began to entrench themselves behind barbed wire and frontline emplacements on both sides in France and Russia. The immediate achievements of the signals war were at sea, but the major successes came later when diplomatic intercepts by Britain brought America into the war.

The foundations and structure of every combatant country's intercept service were laid early in the First World War, along with fault lines in organisation that moulded their performance during the Second World War.

Germany's Allies

Austria made valuable contributions as she had much to teach Germany on the subject of signals interception and intelligence. She had an early start over the other countries in the Central Powers, as well as the Allies, in the development of cryptographic capabilities. A diplomatic crisis arose between Austria and Italy in 1908, during which the Austrians learned how to decode Italian diplomatic radio traffic. This gave the Austrian government, through their *Evidenzbureau* (directorate of military intelligence) in Vienna, an important advantage in negotiations. Further opportunities for practising the bureau's skills occurred when war broke out between Italy and Turkey, and details of the Italian Army's actions in Libya were intercepted by the Austrians as they occurred. As a result, they had one of the best and most experienced cryptographic services in Europe, on a par with the French *Deuxième Bureau*, when war broke out in 1914. Austrian signals intelligence skills enabled them to intercept detailed descriptions of Italian Army plans, although their situation was considerably helped when one of their spies managed to steal the Italian General Staff's communication code. The Austrians, therefore, were more than a match for Italian intercept and cryptology methods, and constantly had the upper hand in reading ciphered messages from the Italian Army. The two armies rapidly became engaged in an entrenched war on both sides, due mainly to Austrian generals misusing the invaluable signals intelligence with which they were presented. It took the Austrians until 1916 before they planned a flank attack to out-manoeuvre the Italian Army with the promised aid of several German divisions to support them, though the German reinforcements did not arrive as the Battle of Verdun had proved more costly in men and materiel than they expected. To compensate for this shortcoming, a major deception was planned by the head of the *Evidenzbureau*, General Ronge, using the skill of his bureau's experienced and imaginative intercept service. He would deceive the Italians into thinking that the German troops that had been promised would arrive to reinforce his forces facing the Italians. The wireless operation that followed was a well-planned covert scheme of complexity, imagination and considerable scale, backed up by the carefully orchestrated appearance of some German troops. These were few in number but made highly visible to the Italians who kept strong reserves of troops out of the ensuing fighting, waiting for an Austrian attack which never materialised. The deception did not benefit the Austrian generals, however, due to their unimaginative lack of follow up in not pressing home their military advantage. This was another demonstration that signals intelligence was only as good as the generals and forces who deployed it in the field.

The Italians, on the other hand, were more experienced in interception in another communications medium – the telephone. The Austrians habitually used the telephone in their preparations to attack the Italian Army, but the Italians tapped into their telephone network and knew where and when the attacks would take place. The Italians then moved their troops back from their frontline defences shortly before the Austrians delivered a massive artillery barrage. They had withdrawn to a second line of defence so that when the Austrians advanced and overran the deserted frontline positions with ease, they then found a prepared second line of defence from which they were repulsed with heavy losses. The battles were bitterly fought on this front, but they did little to influence the war's outcome, largely because of the poor quality of the commanders in the field on both sides. The excellence of Austrian signals intelligence gave them an advantage, but good intelligence does not necessarily turn a poor commander into a good one. The Turks were more effective as allies to the Germans as they engaged the British in Palestine and the Dardanelles quite strongly, though they too appeared to use little signals intelligence in these theatres.

Allied Intelligence

The burgeoning and very effective cryptographic bureau at the Admiralty in Whitehall was housed in Room 40 in the Admiralty and happened almost by accident, with civilians intercepting German wireless transmissions. Cryptographers were almost casually found to start an effective decryption department, although it had its shortcomings. The security of its findings was considered more important than informing commanders of what they needed to know about the enemy. In the first years of the war, senior naval officers played their cards close to their chests so that security somewhat overruled effective use of intelligence. Later in the war, however, the British improved their dissemination of intelligence evaluations and this improvement was carried through into the Second World War, although in the early stages senior officers were still suspicious of intelligence and particularly if it came by wireless. The old guard who were most dismissive of signals intelligence gradually made way for new commanders with a better understanding of intelligence evaluations. The new boys had learned how to use intelligence guidance more widely to help plan major operational decisions, particularly at sea.

The French *Deuxième Bureau* was the best prepared of all the Allied intelligence agencies at the outbreak of war in 1914, and so they needed to be. Many people think of the *Deuxième Bureau* as the French Secret Service, but this is not true. It is the centralised bureau that created intelligence evaluations from other sources, mainly agents. It received signals intelligence feeds from the *Service de Renseignements*, which is still the real French Secret Service, and other sources such as counter-espionage organisations which have their own intercept services. Collectively, the French know them as 'special services'. In peacetime they keep a low profile, as do most similar organisations in other countries, but the *Bureau* emerges as a powerful agent of French interests when danger threatens. What the Germans did with signals intercepts to devastating effect to the Russians on the Eastern Front, the French were also about to do in the west to the German invaders, although they also used spies to good effect. The big problem for Allied agents at that time was to get the information they had gleaned behind German lines into Allied hands, as there was no wireless transmitter available to them as there was in the Second World War. The preferred route for couriers carrying information was to cross the border from Occupied France or Belgium into neutral Holland, but the Germans were aware of this. Their counter-espionage agencies erected electrified fences along the Dutch borders and constantly patrolled it.

Alice Dubois was a French courier who regularly smuggled documents and other agents into and sometimes out of Holland. Much of the information being sent was very detailed data about the frequency and movements of troop and munitions trains, as transport then was almost entirely by train. Agents used secret inks on papers they hid in the lining of their clothes. One of the inks Alice used was made up of lemon juice and onion juice that made no mark visible to the naked eye until the paper was heated. She often got her information across the border by crossing a canal, using a baker's kneading trough as a life raft as she could not swim.

Another agent was Marthe McKenna, the daughter of a Belgian farmer. The Germans thought her father was working in the resistance movement so they burned his farm house down with him inside it. As a result, Marthe became a dedicated agent and used the cover of nursing wounded German soldiers to obtain information for her espionage work and went on to establish a network of fellow agents. Clive Granville was a British agent educated in Germany; he was able to join the German Army, where he rose to the rank of captain, and was able to feed Marthe much useful intelligence. Marthe was given away by the loss of a watch engraved with her initials when she was helping to blow up a munitions magazine. German counter-intelligence advertised the watch as having been found and when Marthe went to claim it, breaching her security code, she was arrested and convicted. Winston Churchill wrote of her in a minute to the Cabinet: 'By all the laws of war her life was forfeit. She did not dispute the justice of her fate.' Marthe died in prison as the war ended.

Many agents worked behind the lines on the Western Front in the First World War, some of the most effective ones being women, but they all needed to find a channel of communications across the border into Holland and on to London or Paris. Major John Oppenheim was head of British intelligence based in Rotterdam and in charge of British espionage networks in Belgium and Northern France. His great concern was getting his agents' reports through the frontier barrier by land, sea or even air. One of his main methods was to hide secret reports in the coffins of the dead whose relatives wanted them to be buried in Dutch soil. Another of Oppenheim's responsibilities was placing his agents behind enemy lines in France or Belgium, often by parachute in a similar way to that done in the Second World War. David Bloch was a French agent who was ordered to collect intelligence about enemy troop movements in Lille and deliver his reports by carrier pigeon. He was given a crate of them to carry his messages but unfortunately for him the large basket of cooing pigeons he had to carry around and feed twice a day made him a somewhat suspicious character. He was captured in his home town, convicted as a spy and shot as were many other brave (and sometimes foolish) men and women. The life of Belgian, French and British agents in the resistance movement behind enemy lines in the First World War was even more dangerous than in the Second World War without a radio to transmit their messages.

4

Europe's War

The Reasons for War

The First World War began with all future belligerents being driven to fight by their own strong motives and convinced that they would win important concessions. War became inevitable. French national pride had been badly hurt when she had been defeated by the Prussians in 1872; as a result of which Alsace-Lorraine had been taken from them. They felt unable to achieve its restoration without help from allies, so when they signed the Franco-Russian Treaty of 1894 it led them to believe that they could get aid from their Russian friends in regaining their lost territory. The Russian motive for war was her need to access the Mediterranean Sea through the Dardanelles which had been lost when she had been defeated in the Crimean War and, as a result, the passage of their vessels through the straits had been blocked. Their Imperial Black Sea Fleet was bottled up in the Black Sea with no access to the world's oceans unless the Great Powers could be persuaded to agree free passage through the Dardanelles. Russia's deadly rival, the Austro-Hungarian Empire, saw the Russian access to the Mediterranean as a threat to their ports on the Adriatic coast and were set against permission being granted. As Austria and the Imperial German Empire were close allies, it was widely assumed that any conflict would trigger a German alliance, bringing them in on the side of the Austro-Hungarians. Russia did not want war with Germany, but the treaty guaranteeing that France would come to her aid militarily in the event of war changed her attitude to all that. Britain did not care much about what happened in Europe, but she did care about Britannia ruling the waves and Germany was building a large navy to protect her trade routes.

Germany was a nation in the ascendancy at the beginning of the nineteenth century, with a population that had increased from 40 million to almost 70 million in twenty-five years. A third of that population was under 15 years of age, which explains how she was able to find such reserves of men to fight the war. The growth in her economy matched that of her population as German exports exceeded those of Britain, France and America put together. The need for a navy to protect trade routes became clear after the Royal Navy arrested two German merchant ships off the coast of South Africa. The German public were incensed and the government started to build a navy which would have a ratio of two ships to every three in the British fleet. Churchill, as a Liberal politician at the time, wrote that Britain's national security was not threatened by the expansion of a Imperial German Navy and that claims of the magnitude of the plans to do so were exaggerated (he changed his mind as he joined the Conservative Party a few years later). The British public, who had been brought up on the view that the Royal Navy was supreme, felt that any change in the balance of power was a threat to their nation.

Militarily, France felt the same way about her army, whose reputation and size they had been building after its defeat by the Prussians almost half a century before. As diplomatic pressures increased, the last thing that Germany wanted was to fight both France and Russia on two fronts at once, but both of those nations were rearming so fast that the German High Command felt increasingly threatened. They pressed the Kaiser to start a pre-emptive war while their army was still strong enough to win it. Meanwhile, Britain had signed an agreement with France that if she were attacked a British expeditionary force would come to her aid. This was a war waiting to happen and all it needed was an excuse. That excuse occurred on the other side of Europe in Sarajevo with the assassination of the Archduke Franz Ferdinand, the heir to the Austro-Hungarian throne, by Bosnian Serb students. The resulting conflict between Austro-Hungary and Serbia started a complicated chain of events, which in turn caused their allies to mobilise according to their treaty obligations. Soon, the whole of Europe was at war.

Europe's armies had emerged from a Napoleonic form of warfare at the turn of the century into one that was to be changed by new and more deadly weapons, although no one realised it at first, least of all the generals. Their mindsets were firmly fixed on the way that they had fought previous wars, where French infantry still wore bright red trousers, making them excellent targets for sharpshooters carrying the new breech-loading rifles. Machine guns were just beginning to make an appearance, although many of the officers were unfamiliar with their workings and thought them unsporting. Then there were those new-fangled wireless sets which were mainly to be found in the staff establishments of generals who probably were not entirely clear as to their purpose. There were men behind the lines, however, who did know the purpose of signals communication systems and had been practising with this new and still emerging technology for some time. Well before the war started, the *Deuxième Bureau de l'Etat* was set up by the French General Staff, who had already set up a desk in Paris to intercept foreign army and diplomatic wireless transmissions. The brief of the *Bureau* was to develop its cryptographic skills by intercepting and decrypting both military and diplomatic signals traffic, so when war broke out the *Bureau* was ready to play its part from the outset.

The *Deuxième Bureau* was one of the first cryptographic agencies of any note in Europe and probably in the world, on a par with the Austro-Hungarian *Evidenzbureau* in Vienna. Austria's cryptographic skills started with their police force, as the institution's name implies, but its military possibilities became clear very quickly; France followed the same route as war began to loom. The French gained an insight into German attitudes by breaking into the German diplomatic code, thus realising the importance of this new technology and beginning to make themselves familiar with the different types of radio traffic. They intercepted as much as they could, to see what they could decipher on a day-to-day basis to increase their skills. Both countries' services now primarily had military intent, but as the name of the Austro-Hungarian bureau hints, they already had some cryptographic experience in telegraphy in police and civil order matters. Operators in many countries soon began to come to Vienna to be trained to recognise differing Morse key styles, use of ciphers, methods of camouflage of transmitters and other techniques of signal usage. The Austrians improved their skills by intercepting foreign transmissions, particularly those of Italy and Germany, well before the outbreak of war. By 1914 the French Foreign Ministry was regularly reading diplomatic messages between the German Foreign Office in Berlin and the German ambassador in Paris. When the German declaration of war on France was intercepted by the *Bureau*, they were able to scramble some passages in its long and involved text, ensuring that the German ambassador could not read it properly. It took time to clarify the contents, mainly by telephone (also tapped), before the document was delivered to the French government. The declaration and its terms were therefore no surprise to the French as the intercept had given them time to prepare a reaction. As this episode shows, the French General Staff were much better prepared for the electronic signals war than the Russians or even the Germans at the beginning of the First World War.

The German General Staff had considered the possibilities of a bureau engaged in cryptography before the war, but did little about it. In contrast, the Russian

Imperial Army had issued wireless sets to units in great numbers, but unfortunately for them they had not paid any attention to operator training or signals security matters, as with most combatants in the coming war, but a sharp lesson was waiting for them. The British Army had made some attempts at setting up an intercept service, but were restricted by lack of funds and lack of acceptance by the Officer Corps of this strange new practice of communications, so they too were about to learn the hard way.

The Battle of Tannenberg

The German High Command's nightmare was to be involved in a war on two fronts, but that is precisely what happened in 1914: they faced the Russians on their borders in the east and the French and British in the west. The majority of Germany's troops were committed to the Schlieffen Plan, an offensive formulated by General Schlieffen, which planned to attack France by passing through Belgium. The German strategy was to hold back the Imperial Russian Army using defensive tactics while seeking a swift decision over the French and British in the west. The High Command estimated that the Russians would need about six weeks to mobilise their immense army, but the Russians surprised everyone by going into action sooner than expected. They crossed the border and advanced well into East Prussia within days of war being declared and the way to Berlin was open. An early Russian victory could have knocked Germany out of the war within weeks and the implications of this opportunity were immense; if the Imperial Russian Army had advanced on Berlin then Germany would have had to admit defeat. There would not have been the dreadful slaughter and suffering of the next four years and it is even possible that the Second World War would not have happened.

The Imperial Russian Army had 416,000 men and almost 1,300 guns in the field, twice the size of the German Eighth Army facing them, which mustered 166,000 men and 846 artillery pieces, commanded by General von Prittwitz und Gaffron. The Russians were divided into two armies: the First Army with 210,000 men was commanded by General von Rennenkampf; and the Second Army with 206,000 men by cavalry General Samsonov. It was said that these two officers disliked each other but, for whatever reason, the fact is that they did not work well together. Prittwitz had earlier fought the indecisive Battle of Gumbinnen against the advancing Russians, the outcome of which was to position the armies in a way that would decide the course of the future campaign. After the battle, Prittwitz exuded defeat and talked of retreat to Berlin in long telephone calls to his superior General von Moltke in his headquarters in Koblenz, over 1,000km away. The telephone network that the German government had installed a few years before was going to repay them handsomely. Prittwitz was immediately sacked by telephone. A few years earlier it would have taken a messenger on horseback a week to deliver the news by a written despatch. A further telegraph message to 66-year-old General von Hindenburg brought him out of retirement to lead the Eighth Army; he famously replied 'Am Ready' and immediately took command. He appointed von Ludendorff as his chief of staff and made plans to fight the battle in a place called Tannenberg. The battle was to prove the first military engagement where the emerging radio technology would play a decisive role. This was remarkable because the German intercept service hardly existed, but signals intelligence gave Hindenburg a decisive advantage just by listening to Russian radio traffic.

Two Russian armies crossed the German/Russian border and pressed on past the Masurian Lakes, which consisted of a series of stretches of water and marshes just north of Warsaw; the lakes separated the two armies, with First Army passing to the north of these watery obstacles and Second Army to the south. Hindenburg decided that he would attack Samsonov in the south if Rennenkampf's First Army, some distance away in the north, was not in a position to attack his flank. The Germans had five of the new mobile radio stations for their army, but it was the heavy ones in their fortresses of Koenigsberg, Graudenz and Thorn that were going to count. There was little transmission traffic in the fortresses to keep the operators busy so they listened to Russian transmissions to pass the time. The Imperial Russian Army was equipped with the most up-to-date wireless sets which were attached to every large formation of their army. Unfortunately for them, their operators were not well trained; in fact, not trained at all. The possibility that their messages could be heard by the enemy probably had not occurred to them so they sent virtually all their messages to each other in clear text. A Russian message was intercepted by Fortress Thorn indicating that Rennenkampf would not pursue retreating German forces to his front. On the initiative of the commandant Colonel Nicolai, the message was immediately despatched by motorcycle to Hindenburg. This was probably the world's first electronic interception of a military signals message in battle conditions and it was an important one. As a result, Hindenburg decided to let the whole weight of his Eighth Army fall on Samsonov's Second Army. The value to the Germans of this intercepted message was incalculable.

As battle intensified the Russians sent all their radio communications in plain text, not appreciating the finer points of radio security. They were to pay a heavy price, as the Russian General Danilov wrote in his report after the battle on their communications failures: 'The use of radio was entirely new and therefore unfamiliar to our staffs. The enemy, however, were also guilty of some errors, although this does not absolve us from the charge of unpardonable negligence.' Danilov considered that faulty functioning and, in most cases, non-existent use of radio security was the major cause for the Imperial Russian Army's catastrophic defeat. The Russians became aware of some of the ways wireless communications could be used later, but they did not appreciate the importance of the maintenance of a dependable radio network until the beginning of the Second World War.

The Battle of Tannenberg began on 23 August as the Russians attacked German positions and several radiograms were intercepted revealing the Russian objectives, including their unit strengths and line of march. Two of the most important interceptions on the night of 24 and 25 August 1914 still reside in the Bundesarchiv and read:

To the Corps Commander of XV Corps
Your Corps will deploy along the Komusing-Lykusen-Persing line till 0900, at which time attack is desired. I shall be in Jablonica, Kljujew with XIII Corps

Then to the 2nd Army Chief of Staff
The XIII Corps will go to the support of General Martos XV Corps and deploy along the flank and rear of the enemy at 0900

This information enabled German generals to avoid an encirclement of part of their army. On the next day of the battle, 24 August, Hindenburg was handed another intercepted message containing the complete operational order of First Army, giving their objectives and timetable. A further intercept indicated that the Russians could not reach those objectives until the 26th, so that told Hindenburg that the Russian First Army would not be able to attack the flank of his army for a couple of days. This gave him time to deal with the Russian Second Army passing to the south of the Masurian Lakes. He moved his headquarters southwards and, on his way, received another intercepted message on the 25th, still in plain text, with details of the organisation and destination of the Second Army. It also contained a somewhat garbled order from General Samsonov to a corps of the Second Army. The text omissions we represent as '…' so the text reads:

After battling along the front of the XV Corps the enemy retreated on 24th in the direction of Osterode. According to information … the land defence brigade …The 1st Army pursues the enemy further, who retreats to Konigsburg-Rastenburg. On 25th August the 2nd Army proceeds to Allenstein-Osterode line; the main strength of the army corps occupies; XIII Corps the Gimmendorf-Kirken line; the XV Corps Nadrau-Paulsgut; the XIII Corps Michalken-Gr. Gardiene, boundaries between the army corps on advance; between XII and XV the Maschaken-Schwedrich line; between XV and XXIII, the Neidenburg-Wittigwalde line. The 1st Corps to remain in District 5, to protect the army's left flank … The VI Corps to advance to the region of Bischofsburg-Rothfliess to protect the right flank. To protect station Rastenburg the 4th Cav. Div, subordinates to VI Army Corps will remain in Sensburg to observe region between Rastenburg-Bartenstein line and Seeburg-Heilsburg line. The 6th and 15th Cav. Div … staff quarters in Ostrolenka.

Hindenburg now knew the order of battle of both Imperial Russian Army's and the other intercepts gave him their objectives.

Another aspect of the battle determined by the interception of plain text transmissions was the manoeuvring and engagement of the Russian XIII Corps of the Second Army at Allstein. Hindenburg was completely in the dark as to this unit's position and its intentions, but he was concerned that it represented a threat to the flank of his own Eighth Army. During the morning of the 28th, further intercepted radiograms describing the Russian XIII Corps' movements and intentions were received. In an innovative move, orders were despatched by airplane to Hindenburg's reserve forces to counter Russian moves. Further intercepts allowed him to feel safe in concentrating German forces on the destruction of the Russian's Second Army. Hindenburg's victory was one of the most complete in military history, with 78,000 Russian soldiers killed or wounded against 5,000 Germans killed and 7,000 wounded. The Germans took 500 guns and 92,000 prisoners. This savage blow kept the Imperial Russian Army off balance until the spring of 1915 and probably accelerated the Russian Revolution. Meanwhile, Rennenkampf's army had gone on the defensive until Hindenburg attacked him, so consequently the Russians drew back beyond their own borders with very heavy losses.

One of Hindenburg's last strokes in the campaign was a deception using radio transmissions. During the Masurian Lakes campaign later in September, he needed to mislead the Russians into tying down their large reserve force around the city and fortress of Koenigsberg. Hindenburg had no extra troops available to engage this reserve force so he decided on a tactic of misinformation. On 7 September the radio station at Koenigsberg sent a purposely disjointed message:

To the Corps Chief, Priority Telegram
Guard Corps

Tomorrow the Guard Corps will join the ... Immediately west of Labiau, parts
of the V army unloaded

Army Staff Headquarters.

The Russians intercepted this deliberately plain text message and were completely
taken in by it. The Russian reserves did not move to support their comrades who
were under attack, but waited patiently for an attack from a German formation
that did not exist. This is one of the first recorded 'double cross' messages that
became an art form later in the war and particularly in the Second World War.
German operators began to recognise the style of Russian transmitters, but
were put under orders not to jam them on the principle that a poorly secured
transmitter was of such value as an intelligence source that they should be given a
clear field. As Napoleon said, 'never interfere with the enemy when he is making
a mistake'.

The Battle of Tannenberg and the following campaign were shaped by a number
of things. The Russians crossed the border into East Prussia just a few days after
war was declared. They were able to move quickly because they had not taken the
time to prepare the logistic supplies and neglected the needs of their armies in
the field. Also the German Army was of superior quality to the Russians, having
better equipment, training and leadership. Rapid movements using Prussia's rail
system enabled the German staff to entrain and move complete corps formations
around the battlefield and concentrate them against threats or opportunities
posed by the enemy. Russian signals transmissions consisting of over a hundred
messages in plain text had enabled Hindenburg to know the enemy's position,
strength and intentions days in advance of their being committed to action. None
of these advantages would have been worthwhile if the German commander
had not been a superior general who was able to inflict grievous damage on the
Russians at little cost to himself. His victory virtually cancelled out the threat of
the Imperial Russian Army on Germany's eastern border, enabling the German
High Command to concentrate on the battle in France. However, on reflecting
on the battle, Hindenburg said almost nothing about the intercepted messages
from which he had benefited.

SIGINT in Galicia

Meanwhile, on a different front in Galicia, south-eastern Poland, the Russians were engaged with the Austro-Hungarian Army commanded by General Holtzendorff, who was defending the threat of an advance on Vienna. The Austrian High Command in 1916 became aware from signals intercepts that the Russians were getting good information about their troop movements and, particularly, the weak points in their defence. Major Nowotny, a counter-intelligence specialist, was ordered to investigate the leaking information, so he visited many units at the foot of the Beskids Mountains in a part of the front he was convinced was the source of the leak. One unit he visited was a field hospital run by the Sisters of Charity whose reputation for tenderness and compassion was loudly expressed among the Austrian wounded. Outstanding among them was Sister St fInnocence, who comforted the men and made them laugh at stories about her convent in Vienna while she helped them to write letters home.

The major interviewed all the staff of the tented hospital and closely studied their identity papers which were without fault. Further enquiries found Sister St Innocence to be so outstanding among the nurses that she had been recommended for the Red Cross Gold Medal for spending so much of her off-duty time among the wounded. Nowotny interviewed her along with other nursing staff and was struck by the size of her feet peeping out from under her long nun's habit. He pretended to faint and fell into her arms, but just as suddenly he recovered to order the Sister to be stripped and found that 'she' was quite a muscular man. The major demanded his name. 'I am Captain Vassily Vassilijevich Gerson on the staff of General Dimitrov of the Russian Imperial Army,' he replied.

Meanwhile, Holtzendorff continued to retreat in disarray before the Imperial Russian Army and was forced to call upon the German High Command for reinforcements. The newly formed Ninth Army arrived after some delay as Hindenburg had only just completed his victory at Tannenberg, bringing with them knowledge of the uncoded intercepts of the Russian Army. By this time the Russians had learned some lessons with regard to transmissions security and had begun enciphering all messages. Unfortunately for them, the Austrians were experienced code breakers and had no difficulty in reading their rather naive transmission codes. After a series of actions in which the Austrians read much of Russian radio transmissions, the fighting petered out with both sides exhausted. The battle on the Eastern Front was over for quite a time. As for Captain Vassily Gerson, he faced a firing squad the morning after his being unveiled. He had spent five months in a nun's habit and had gleaned much intelligence about Austrian dispositions on that sector of the front. It seems a poignant end to a brave and resourceful soldier who took holy orders to serve his country.

The Miracle of the Marne

On the Western Front, a German attack was made through Belgium as part of the Schlieffen Plan, which was designed to sweep onward through northern France to crush the French Army. This theatre of war became one of rapid movement, but signals intercepts were going to decide much of the battle's outcome as well. The French *Deuxième Bureau* on the Western Front was well prepared for the signals war and were determined that they were going to defeat the German Army's attack, even though they did not have the benefit of the plain text messages that Hindenburg enjoyed reading during his campaign in the east. However, they were able to decipher the German messages quite easily. The Miracle of the Marne was not the miracle it was made out to be as it was primarily due to the French practising their skills in intercepting and deciphering enemy signals well before war began. The German High Command had planned their army's advance through Belgium with a great sweep east around Paris to surround the French Army and destroy it. On their right wing, the First Army was commanded by General von Kluck. His subordinate, General von der Marwitz, commanded a cavalry corps that made good use of their radio equipment in the German fast-moving advance. Von Kluck's rapid advance through Belgium and then into France used radio extensively to co-ordinate the units of his army according to plan. However, German radio operators had little training in signals operations and their skills in the new discipline were slim. They sent transmissions correctly in cipher to begin with, but as the heat of battle increased, messages were sometimes sent in plain text and security procedures began to flag. The French intercept service began monitoring the German Army's radio traffic even before they began to cross the Belgian border, and this enabled them to track regularly and with increasing detail the positions and movements of their advancing enemy.

The *Cabinet Noir* was the cryptographic department of the *Deuxième Bureau* and intercepted over 350 radiograms transmitted by the German cavalry corps over a two-week period during the campaign. The *Bureau* called their interceptions the 'Marwitz Telegrams' as the German wireless operators disguised their call-signs less and less effectively, as the stress of battle made them more lax in their wireless security disciplines. Messages were hurriedly enciphered (or otherwise) by the radio station officers, who often had little understanding of the reasons for them.

Radio station staff had no clear instruction on wireless security so the call-signs of each station in the army invariably started with the same letter and remained unchanged as their advance progressed, nor was there any change in wavelength of the broadcasts. The French were able to establish the wireless stations of every German Army division by their individual call-signs. Cavalry units were the worst offenders, probably due to stress of their fast-moving formations, although some infantry divisions and even corps developed bad security habits as well. Each German cavalry control station, for instance, had an identifying letter: 'S' was the designation of units in Belgium, 'G' in Luxembourg, 'L' in the Woëvre and 'D' in Lorraine. Confirmation from some messages came in plain text and could even be clearly signed by the sender with their rank and name. After a few intercepts, it became known that General von der Marwitz commanded the corps using the 'S' letter in Belgium and General Richthofen commanded the corps using the letter 'G' in Luxembourg. A clear message with a call-sign 'L' stated that two cavalry divisions had forced their way into the Woëvre Valley and were moving towards Verdun via Malavillers and Xivry-Circourt. This kind of information was extremely valuable to the French general directing his battle. After a few days of these interceptions, the *Deuxième Bureau* were able to describe to the French General Staff the operational structure of the enemy forces they were facing in detail. The *Bureau* followed the movements of von Kluck's First Army as it advanced through Belgium and from this were also able to extrapolate and deduct the structure and strength of Second Army under General Otto von Bulow. These two armies were unable to keep in touch with each other as they wheeled in a great arc across France and a widening gap began to appear between them.

Von der Marwitz's cavalry were ordered by radio to provide a thin screen of lancers to cover the widening gap between First and Second armies. The French identified this as a weak spot in the German front that began to stretch for miles as the two armies advanced at an uneven pace. Using signals intelligence gleaned by the *Deuxième Bureau* on 8 September, the French general struck at the critical point between the two German armies' line of advance. They soon began to threaten the German First Army with encirclement and outflank von Bulow's Second Army in the process, causing both German armies to retreat. The German High Command was blamed for ordering the retirement when the Battle of the Marne was 'almost' won in the minds of the German public. Frontline French soldiers were surprised by the change; the German Army retreating in the face of a desperate French resistance became known as 'The Miracle' in public parlance. The French High Command and the *Deuxième Bureau*, however, knew better.

The Race to the Sea

The British had formed an Army Signal Service in 1912 as part of the Royal Engineers at a time when there was little money or resources available from the War Office. They did not envisage the size or complexity of the conflict that was to come, so wireless communications were not a priority and were rarely used by staff officers who probably did not fully understand, and even mistrusted, the new-fangled codes and ciphers. The intercept services of the British Expeditionary Force (BEF) in France were, therefore, under-used in the war's opening stages so operators had time to listen to enemy transmissions in a similar way that the German operators had done in Thorn Fortress on the Eastern Front. Operators began to intercept enemy transmissions and gauge the intentions of the enemy, so transmissions were restricted or being sent by the more traditional means of runners or riders to keep the airwaves clear. It is unlikely that they used Hindenburg's innovation of sending despatches by aeroplane, but one novelty that the BEF did have was a 'wireless compass' device which was issued to them in 1914. This was a direction-finding receiver made by Marconi and further developed by Professor Sir John Fleming (who coined the word 'electronic'). The newly instituted Direction Finding Service started using its new equipment in locating enemy transmitters for the BEF, and this was later taken up with good effect by the Royal Navy. Radio stations were invariably attached to the headquarters of an army formation and it was possible to locate German formations by 'wireless compass' as the message was being transmitted. The ability to spot the source of enemy transmissions and thus piece together their order of battle became increasingly important during the course of the war.

General von Kluck found himself checkmated at the River Marne, north of Paris, so he switched his attack further west to outflank the French, British and some Belgian forces that were positioned nearer to the French coast. The Germans probed the French and British line of defence in the direction of the Channel in a deadly dance of men and guns while trying to find a vulnerable flank. Massive German troop movements were constantly monitored by the *Deuxième Bureau* and the less experienced intercept service of the BEF as the opposing armies manoeuvred around each other.

The previous experience of British military intelligence had been formed during the Boer War some fifteen years earlier. Even at that time they had realised that the gathering, analysis and use of intelligence information needed method and had developed a three-tier reporting structure: intelligence officers collected information from frontline troops; this was sent on to staff officers to be collated; and the results were then analysed by a Field Intelligence Department to assess the enemy's strength and intentions. This proved quite successful in action, but the little signals experience that British Army operators had of SIGINT was when the Boers captured some of their few radio transmitters: the main lesson learned was not to leave your wireless sets lying about when the enemy are near. As the war in South Africa ended, the memory of the intelligence structure the army had worked out for itself began to fade. The experience would have been entirely lost but for a manual written in 1904 entitled *Field Intelligence: Its Principles and Practice* by Lieutenant Colonel David Henderson. This document proved invaluable to the War Office in its sudden and unexpected mobilisation in 1914 as they realised that an intelligence system was needed by the BEF within its command structure. This was the new intelligence handling system into which the signals intelligence operators began to feed their intercepts, in addition to those of the *Deuxième Bureau*, to provide British staff officers with a clear picture on the strong German forces in front of the BEF.

After the war, an analysis of German signals intercepts by Colonel Cavel of the French *Deuxième Bureau*, correlating intelligence evaluations with movements of Allied forces, showed that counter-measures and effective actions taken in the 1914 battles of movement were almost all due to signals intelligence. Another of the colonel's findings was how quickly the French and British operators learned to use the skills of electronic warfare to counter the enemy in the early months of the war. The same could not be said of the German signals intelligence, who took almost a year to develop an effective intercept and decryption service. By 1916, however, both sides on the Western Front had developed comparably efficient signals intelligence services as the levels of wireless traffic increased and skills in decryption and security improved. The German Army had learned a lesson about signals security at Tannenberg on the Eastern Front, but the victors were slow to apply that lesson to the security in signals transmissions in the west. The German General Staff showed a lack of awareness of the wireless security faults that had betrayed the Imperial Russian Army and this would now act as a weak point in their own conduct of the war for over a year. The war of movement gave way to a war of entrenched positions along a 350-mile-long front by 1915. A fundamental change in the nature of the signals intelligence war began to evolve as the conflict went into its second year. Trenches and barbed wire entanglements now extended from the Belgian coast, across the fields of France and to the borders of Switzerland. No major change in those emplacements would occur in the years to come until the war of movement began again in 1918.

In the Trenches

The war of movement slowed to a stop as 1915 dawned and one of stagnation developed, with the static battle lines of positional warfare drawn firmly on the map of Europe. Entrenched armies faced each other across a no man's land lined by aprons of barbed wire, swept by machine-guns and churned up by artillery, forcing defenders in their trenches to think of ways to outwit their opponents. The nature of SIGINT had to be modified to suit the changing needs of soldiers seeking to discover their opponents' intentions. A dense criss-cross of telephone wires for telegraph and field telephones became the principal means of communication between infantry and artillery units for receiving orders and reporting results of actions. Wireless telegraphy began to be used for the purposes of spotting the fall of artillery shells from biplanes that had not yet learned to shoot at each other with properly mounted machine-guns, although aerial photography was also evolving as a technique to confirm or deny intelligence reports. Wireless communications in close-quarter ground fighting decreased in value until the war of movement was resumed three years later. Portable wireless sets were a rare commodity on both sides until 1916; the one that British troops had to deal with was the British Field Trench Set. It needed six or more men to carry the set and the heavy accumulators that were vulnerable to damage or spillage of acid. The tall aerial also made it an excellent target for the enemy, so the whole contraption was far from suitable for use in forward areas of the front. The Loop Set, which began to be issued in 1917, was more portable and had a 3 sq ft aerial; unfortunately it only had a limited range of about 2,000 yards. The spoken word on the field telephone took over from Morse code as the means of communication for the entrenched armies.

The telephone also had its drawbacks, of course, but the stabilisation of the fronts made the network a more effective means of communication than wireless. A cat's cradle of wires and cables covered the territory behind the lines, but they were easily and often broken by artillery fire. Early in the war it was found that cables needed to be buried 6ft under a protective metal shield to survive heavy shelling; however, water in the soil could also short them and could make a connection a hit-and-miss affair. Considerable trouble was experienced from cross-talk (i.e., interference from another transmission), so that when an officer picked up a telephone, either in headquarters or the trench dugouts, he was never sure to whom he would be talking. At the same time, intelligence officers became disturbed about the leakage of tactical information and how well informed the enemy was. Routine operations in the trenches, in particular British ones, could attract well-directed artillery or machine-gun fire for no foreseeable reason. Units which moved into trenches to relieve other formations would be greeted with welcoming notices chalked on boards held up by German soldiers naming the regiment of new comers; in one case they even named the commanding officer. French units began to report similar experiences. Spy scares were rife.

Could the cross-talk that was being experienced by everyone have anything to do with a leakage of information on the field telephone network? The Signal Service was told to investigate; they knew that electro-magnetic energy could be transmitted through space, but could it be transmitted through earth as well? Scientists began to experiment; German scientists such as Ohm, Gauss, Kenz and Hertz had been prominent in the development of electro-mechanical technology and wireless transmission, but could they have developed a devilish scheme? Then a report came from the French Signals Service that there had been attempts by the Germans to tap into French artillery telephones by trailing wire into the bed of a stream running through the French positions. The source of the leak of information became clear. There were railway lines, pipelines and watercourses running across the positions between combatants, and these carried the electro-mechanical signals that enabled the Germans to intercept and 'overhear' conversations of Allied officers quite clearly. Intercept techniques now took a different turn and both sides learned how to listen to the telephone conversations of each other.

In 1915, near Apremont, the French were planning an attack and listening to German telephone conversations; what they did not suspect was that the Germans were also listening to them. The Germans learned what time the attack was due and its objectives, so they passed the information on by telephone to their own troops. Ironically, it was by telephone that they originally got the information and, as a consequence, the French tapped into the conversation and put forward the time of their attack by several hours, which caught their enemy by surprise just as they were rearranging their troop dispositions.

An ageing German cryptographer in Berlin described to the author what he thought of as a major achievement of the German intercept service during 1917. Concentrated and intense levels of traffic in voice and wireless transmissions were indicating a build up for a major attack from the Allied side. The objectives and timing of the attack could be clearly seen by the German High Command from overheard radio transmissions and telephone calls. The Germans withdrew their infantry from forward trenches to prepared fortifications, in what later became known as the Siegfried Line, just before the assault was due. This left Allied troops to assault territory which the enemy had vacated, so their artillery bombardment and other preparations did no damage to the enemy or his equipment, and the attack petered out into empty space. My informant alleged that the Allies had intercepted German radio traffic and, although they realised that they were retreating, did not continue the advance to follow up the enemy withdrawal.

Telephone security was not strict in 1915 on the battlefield, mainly because the change from Morse to the more familiar conversation form on a telephone created a more relaxed situation as one soldier talked to another. In addition, most officers and NCOs who used the telephone did not have the same security awareness as the wireless operators who were trained to observe caution. Unfortunately, the land-line system leaked like a sieve and it was some time before it became generally known to frontline troops.

The Direction Finding Service

Reading the enemy's messages was one thing, being able to locate the transmitter would prove to be another. The BEF 'wireless compass' unit was soon being developed into a fully fledged Direction Finding Service. Using these listening devices enabled an operator to fix the direction of an operating transmitter; two or three 'fixes' on a compass bearing from different D/F stations could give operators a simple plot on their map of the extended bearing; and where two bearing lines crossed would plot the location of the enemy transmitting station, sometimes with an accuracy of 100 yards or so. D/F was largely a British innovation and the German Army did not develop their own version until almost a year later. Wireless technology of the period was still in its infancy and the resources to develop it were scarce. The maps used to plot the transmissions did not carry local magnetic deviation data or even accurate elevations of the land, as they do today, so operators had to learn the effect that terrain would have on D/F readings of enemy radio transmissions by experience. A high degree of skill and persistence was needed by operators to plot the location of the transmitter, so he probably became a valued if junior member of the headquarters staff of the unit. The two disciplines of signals interception and direction-finding became partners in the signals war.

With growing experience of signals interception, all armies began to practise stricter discipline and secrecy in signals transmission, not only with the content of messages but also the call-signs of their transmitter stations. Then as direction-finding techniques improved, the disguising of station call-signs became less important, although call-signs could still give some intelligence information. The geographical location of an identified enemy wireless station could give away a position and even the movements of an army unit, so D/F quickly became an integral part of the signals contest. Tracing a unit's movements by direction-finding techniques was particularly effective on the Eastern Front, where the geography over which combatants were fighting made it a more fluid war than the one in France. The Italians used D/F techniques very effectively during the Austro-Hungarian Grand Offensive in 1917. It was used to some extent by the Russians as they supported the Romanian Army, but they never did get the hang of D/F or even the practice of cryptology in general by the time the Soviet government signed a peace agreement with the German government in 1918.

The French Goniometric Service

The French Army proved fast learners in the D/F field and soon set up their own effective direction-finding or, as they called it, Goniometric (from goniometer, an instrument used to measure an angle) Service, probably in co-operation with the British. Collaboration between the two armies and governments over cryptographic techniques appeared to be close throughout the war as they had much to learn from each other. The same could not be said of their treasured code or cipher keys, though, as those secrets were more hardly won and more easily lost so no one trusted anyone else's security. The French Army developed a mobile direction-finder with a rotating aerial frame mounted on a truck and this innovation gave them very good results. Their goniometric skills and equipment proved very worthwhile later in the war as the Germans prepared to retire to new defensive positions in the Hindenburg Line. Long before withdrawal began, French and British radio stations were set up to monitor activity in the newly established German line of defence that was under construction. The radio stations that the Germans were locating in the new defences were immediately recognised and located by the French, so the exact positions of emplacements were clear to the French using their Goniometric Service. Their intercept service also established when the enemy planned to retire, so a fleet of the new mobile D/F units patrolled the entire front. The German intention to retire beyond the Vesle and to Aillette had been established by French radio and Goniometric Service using a large number of direction fixes as well as intercepts. But, just then, German cipher keys suddenly changed and the French could not read their transmissions. Enemy radio stations regularly changed call-signs but could not disguise their position or the movements that units made. In particular, locating the German Army weather reporting stations moving deep into the rear gave the Allied generals an indication that the German Army was retreating and weakening.

Besides locating wireless stations, co-operation between French and British artillery specialists focused on researching direction-finding techniques for the location of German gun emplacements. Their plotting was most important in the battery and counter-battery duels that were taking place all along the front in France, particularly in preparation for an infantry assault. Variations in methods for spotting an enemy artillery position, including sighting gun flashes and various measurements of aerial photography, dropping smoke bombs on gun emplacements for artillery ranging and sound-ranging of the gun being fired were all being investigated. One of the researchers in sound-ranging was Lieutenant William Bragg, a Nobel Prize winner in co-operation with his father on X-ray research. Incidentally, William's father, Professor Bragg, invented ASDIC (underwater detection device used in the Second World War). Lieutenant Bragg was a part of an Anglo-French team working on sound-ranging with microphones, although they were fairly primitive at this time. Firing a field gun or a heavy artillery piece can also produce a similar 25Hz sound wave, but sounds in this low-frequency range hardly created a response in the microphones that were being used.

Sound-ranging research was not doing well, but inspiration came to Lieutenant Bragg while in the lavatory. The latrine in Bragg's billet was a small closet with no windows and, when the door was closed, the only opening to the air was under the seat. In the seclusion he noticed that the firing of a gun, whose emplacement was nearby, lifted him off the seat with the pressure created by the detonation, even though he could not hear it. So he decided that they should measure the pressure wave created by the German guns. Another problem was experienced in the firing of high-velocity guns, which made two sounds: a loud crack when the shell broke the sound barrier, and a boom that was almost masked by the crack and the 'gun-wave', which could not be used in detecting the gun's position. In addition to all this, a microphone was sensitive to buzzing bees, rifle shots and other noises which blocked out the artillery sounds. Lieutenant Bragg constructed a pilot device made up of used ammunition boxes and platinum wire stretched through them which made a primitive galvanometer, sensing the energy as the gun was fired by the wire quivering.

There were problems yet to overcome, such as the direction of the wind which would blow the pressure waves away from the device. The ideal conditions were in misty weather, with no wind and a consistent temperature. Six or eight of these primitive galvanometers strung out on stakes over a 1,000-yard line facing the enemy made it possible for a skilled operator to identify a particular gun within 100 yards of its location in a battery when all of them were firing. Even the calibre of the gun could be estimated and this would often be confirmed by sending out a patrol to the shell hole to recover the fuse which was sometimes left after the detonation. Sound-ranging was to have a tremendous benefit on British offensives after 1917. German artillery units were issued an order not to fire their gun when the wind was in the east or when the sector was quiet. As French and British troops captured targeted gun emplacements, counter-battery fire was found to be accurate on 90 per cent of their target guns. The effectiveness of sound-ranging and related techniques was demonstrated in the Battle of Cambrai in November 1917 when the British attacked with massed tanks, crushing German barbed wire, and 1,000 guns suppressed German artillery almost completely. The attack stalled as British artillery had to move forward into a sector that had been unmapped for sound-ranging and immediately came under heavy German counter-battery fire. The most outstanding achievement performed by sound-ranging was on Vimy Ridge where a heavy German artillery piece was punishing Canadian troops from a camouflaged position over 10 miles behind the lines. A careful calculation by several sound-ranging units managed to pin-point the target and destroy it. Another major achievement for sound-ranging came in an offensive at Amiens early in August 1918 during a week of dense fog. The Germans used the cloak of fog to move their artillery to fresh emplacements, but sound-ranging located the new positions accurately before the attack began. The German artillery was almost totally destroyed and 450 tanks advanced with a screen of infantry troops on a front about 6 miles long and 10 miles deep. Hindenburg said it was the blackest day for the German Army and it marked the beginning of the war's end.

Romania and Russia

The Russians had been brought to a halt by Hindenburg on the Eastern Front, but they were to try again further south in March 1916 with disastrous consequences. A major attack on a front 200 miles wide was made against the Austro-Hungarians in June 1916 in what became known as the Brusilov Offensive. The attack started well for General Brusilov, whose offensive was probably the most successful of the whole war. His soldiers broke the back of the Austro-Hungarian Army, which lost over 600,000 men, plus 400,000 prisoners. The Russians lost almost a million men themselves in the campaign but could make good the loss over time. The result was that, for the rest of the war, Austria was now largely dependent on its German ally for support in its part in the war.

Another shock for Emperor Franz Joseph came from his country's neighbour and ally, Romania. In 1913 the government of King Charles of Romania renewed a treaty of alliance with the Central Powers of Germany, Austria-Hungary and

Ottoman Turkey. This committed Romania to enter the fray if any nation attacked any member of the Central Powers; in addition, the treaty provided that Italy would also enter a conflict caused by such an attack. As Austria declared war on Serbia rather than was attacked, there was no commitment on Romania's part to join in and she was able to maintain her neutrality under the treaty. Then, in October 1914, King Charles died, and with him ended the close friendship between himself and Emperor Franz Joseph. Romania's government and even her people had begun to lose sympathy with the cause of the Central Powers, but would not declare enmity towards such powerful military neighbours. So they bided their time.

Meanwhile, Austria's ambassador Count Czernin was a diplomat of the old school who believed that your word, or rather your treaty, was your bond. He was unable to perceive the change in attitude of the Romanian government and people so he sent radio reports, using his diplomatic code, back to Vienna telling them of Romania's love of peace. The German ambassador took the same view so they both were totally deceived by the Romanians and suspected nothing of their rising discontent, although the Austrian military attaché saw things more clearly. The Romanians were able to maintain their deceptive measures by reading the radio messages sent by the two ambassadors to their governments at home. The Austrians had shared military cipher codes with what they thought was a friendly nation but not the diplomatic ones. The Romanians maintained their deception in masterful fashion, keeping up their façade of traditional friendship with Austria, due in part to Count Czernin whose lifestyle as an ambassador was an agreeable one. One afternoon he had taken a horse-drawn cab to one of Bucharest's fashionable cafés for a dalliance with a lady or two of his acquaintance when he found that his briefcase had been left in the cab. He immediately notified the police who found it and returned the case and its contents a day or two later, including his diplomatic cipher for communicating with Vienna undisturbed.

Meanwhile, the Austrian military attaché grew increasingly wary of Romania's intentions, particularly as the Russian Brusilov Offensive had begun with such success. In August 1916, the Romanian ambassador in Vienna delivered a declaration of war on their erstwhile allies the Austrians. Simultaneously the Romanian Army marched across the border into the Transylvanian region of Hungary to seize territory they thought traditionally was theirs. The Austrian attaché's fears were shown to be justified. The German government was not included in the declaration but, as Austria's close ally, she immediately became involved, although very much committed to her massive assault on Verdun on the Western Front. The Austro-German intercept service was not monitoring Romanian radio traffic until war was declared and indeed there was little to monitor as transmissions were few and far between. As hostilities started, however, the ether became alive with radio messages and Romania's radio operators transmitted detailed listings of their order of battle which they radioed by fixed call-signs largely in plain text or a code that was child's play for the expert code breakers in the Austrian Army. The mistakes the Romanians made in radio security were worse than those made at Tannenberg by the Russians, and they were taken full advantage of by the German General Falkenhayn in the ensuing battle. In addition, the Brusilov Offensive, which was seen as a support for Romania, slowed to a stop against fierce German resistance and, as it did so, Russian troops infected by revolution began to refuse to take orders from their officers. The results were inevitable and the savage defeat inflicted on the Romanian Army by Falkenhayn prompted the German press headline saying that the judgment of God had fallen on Romania.

The war continued and, a year later, Austrian troops entered Bucharest in a vengeful mood and searched the Foreign Office where photographic copies of Count Czernin's code were found. All this time both military and diplomatic codes had been in the hands of the British for over a year. Not only had the deceptive Romanians been reading the Austrian government's military and diplomatic transmissions, but so had the Allies.

The Austro-Hungarians were growing tired of the war in spring 1917 and the new emperor, Karl I, decided to contact the French with a view to starting secret peace talks. France was keen to see progress in such talks as they had a mutiny in their army and a weakening of Germany's alliance would have been welcome to them. The French Prime Minister Clemenceau asked Major le Comte Armand of the *Deuxième Bureau* to contact his friend Count Nicholas Rovertera, an advisor to the Austrian government, to meet him in neutral Switzerland. The Italians were also involved as their army faced Austro-Hungarian troops on their borders; their terms included taking Trieste from the Austrians as a part of the peace settlement. The new King Karl of Austria found this unacceptable as Trieste was the main port of the Austro-Hungarian Empire, but the Italians insisted on this condition as a part of negotiations. Talks dragged on for almost a year under the direction of 'Tiger' Clemenceau of France, so he set a time limit for their completion. Count Czernin, now prime minister of Austria, made a speech and admitted the talks had taken place. The German government, who had begun to suspect the existence of such negotiations found Armand to be an embarrassment. He was later found shot in his flat in mysterious, and convenient, circumstances for Germany.

The Central Powers were now in retreat and its junior members had long since lost the taste for fighting a losing war. The only battlefield that they now stood any chance of winning was at sea.

5

The War at Sea

From 1914 the conflict at sea fell into two, almost distinct, parts: the surface engagements of the bigger ships, sometimes in line ahead, pounding each other with their heavy guns; and that of a secret U-boat war of sudden and unexpected torpedo attacks. This last part had political as well as economic effects. The Royal Navy's strategy was to use their big guns to enforce a total economic blockade of ships supplying Germany and her allies. The North Sea was to be a war zone and boarding parties from the Royal Navy ships made stringent checks on neutral shipping for any goods meant for a German port. It was a brutal, effective and probably illegal weapon, in response to which Norway, Sweden, the USA, as well as other countries, made loud protests about the contravention of international law. The British did not acknowledge their protest so the Royal Navy maintained a very effective blockade with a relatively small number of warships because of the good intelligence from wireless intercepts about German shipping movements. In response, German U-boats were ordered to sink any ships found in British waters in an act of unrestricted warfare. A savage battle was fought in the First World War, as well as the Second World War, as the two opponents tried to strangle each other to death. The stakes were that the population of the loser would starve. The Royal Navy was well equipped to combat surface ships, but was woefully unprepared for the U-boat threat; however, signals intelligence was going to play a crucial part in the great sea battle that lay ahead.

The Admiralty had fitted wireless telegraphy in all its major vessels by the beginning of the First World War, but was still unprepared for the signals intelligence battle in which they were about to become engaged. They had some hints of what part wireless telegraphy might play from their experience in observing the use of wireless telegraphy in 1904 during the Russo-Japanese War. An ageing coal-fired

Imperial Russian Fleet left their base at St Petersburg to undertake a mammoth trip through the Suez Canal and into South East Asian waters to finally arrive in the Sea of Japan to engage the Imperial Japanese Navy. The British, and indeed the world, watched the slow progress of Russia's fleet by monitoring its wireless communications during the course of their long journey. The W/T specialists on board the Royal Navy ship HMS *Diana*, which shadowed the fleet, commented on the poor standard of transmission and encoding of messages coming from the cumbersome Russian fleet as it chugged halfway around the world to confront the Japanese. The Russian Imperial Fleet engaged the Japanese battleships and was almost totally destroyed, distress calls transmitted by sinking battleships filling the airwaves. Russia's defeat was reported back by wireless to the Admiralty in London, but the lessons of the extended exercise in the use or misuse of wireless telegraphy were ignored. The British Admiralty had no signals intercept service or plan to create one in 1914 as war started, but luck was on their side.

Room 40 at the Admiralty

Within days of the declaration of war in August 1914, the Admiralty's recently appointed Director of Intelligence Rear Admiral Henry Oliver was being given copies of intercepted German wireless transmissions in code that he was unable to read. Winston Churchill, who was First Lord of the Admiralty, and the soon to be appointed First Sea Lord Admiral John (Jackie) Fisher, asked Oliver to set up a wireless intercept service. The Admiralty allocated Room 40 in their Old Building to house this new service. Room 40 would be the title of a signals intelligence bureau that would achieve extraordinary things, first for Britain's war at sea and later as an intelligence centre of wider dimensions. Oliver needed a director for the embryonic intercept service so he turned to his old friend Sir Alfred Ewing, who was Director of Naval Education, and over lunch at the United Services Club, now the Institute of Directors, he offered him the job. Ewing had been a great success at educating navy personnel and had received a knighthood; in addition, Oliver knew that he had an academic interest in codes and ciphers. Ewing, immaculately dressed and conscious of his position and dignity, was an inspired choice for the job with more than his share of luck in its success. Ewing's first move was to review what there was in coding and ciphering expertise in the archives of The British Library, the General Post Office, Lloyds and other repositories of literature and experience on the subject. His conclusion was that he needed a multi-talented team to help him in the task whose shape was beginning to emerge.

German wireless transmission intercepts were coming in thick and fast from radio listening stations, or 'Y' stations as they came to be known, that began to be installed all along England and Scotland's east coasts. Piles of German transmissions in code and a few in plain text from new 'Y' stations were gathering at the Admiralty, in addition to other contributions from radio amateurs who began to play their part in intercepting German messages. Ewing's ever-increasing supply of messages in cipher needed code breakers to work on them, but also people who could understand the purport of a message in naval terms and practice. The intercepts were all in German, too, so the team that Ewing needed to recruit had to contain talented cryptanalysts, naval officers and translators to create a bureau that would help to shape the war to come.

In the early days, Room 40 in the Admiralty Buildings in Whitehall began co-operation with the Army at the War Office just across the road. The army did not have the flying start in putting together a decryption team that the navy was enjoying, but progress with interception was shared. Some decryptions of German coded messages began to emerge as Ewing's small but growing team began to bed down in their task. This co-operation did not last long, however, as Room 40 began to prove itself the better bureau of the two and was rather pleased to show it. In Germany at the same time there was no initiative to create an intercept service and, when a German signals intelligence team was finally inaugurated, it proved trivial compared to the British effort. Even Room 40 was not without its shortcomings, however. Intelligence has several aspects to it, each of which needs to work in balance with the others. The accurate interception of a coded transmission was the first stage. Decoding it was next; a translation was needed, which required a knowledge of German marine terminology and practice to make sense of the message. From those fragments of information, and maybe others collected earlier, an evaluator's skill was needed to build a coherent summary of intelligence for the field commander. These evaluations could support and guide life and death decisions for those directing the ships or ground forces in the face of the enemy.

The directors of intelligence had to decide to whom they should show this precious knowledge of the enemy. Cast the distribution list too narrowly, as Hitler did, and you limit the use of valuable intelligence, but cast it too widely and the risk of a disaster in the shape of a leak could betray your sources to the enemy who would promptly close them down. It was a difficult balance between too open-handed a distribution of your intelligence and paranoia about the enemy's spies that are always a risk to your secret sources and plans. Room 40 as a bureau had limited its effectiveness by restricting its agenda to acting largely as an interception and decoding role rather than a wider intelligence centre into which it would evolve later in the war. Oliver and Ewing had built up an extraordinarily effective decoding facility, but would play what intelligence cards they had very close to their chests. This meant that any useful intelligence their cryptanalysts had generated had a more limited effect than it could have done. Later in the war, Room 40's capability came into its own as a fully effective, intelligence-gathering, evaluation and exploitation organisation under the inspired leadership of Admiral 'Blinker' Hall. He was not only a superb spymaster but a great judge of men; one of his many lieutenants, Commander Alastair Denniston RN, served his signals intelligence apprenticeship under Hall and was to make his mark in intelligence later on.

To meet the need of the unsolved coded intercepts the Admiralty began by recruiting bright young men and their tutors from the universities for cryptanalysis and linguistic work in the Room 40 team. The foundation of Britain's signals intelligence bureau and their cryptanalysis skills was being laid by Ewing's growing team; its effectiveness would last for half a century or more. Sadly, a damaging quarrel between the War Office, or MI 1B as it was designated, and the Admiralty across the road in Whitehall was not resolved until 1915. Such squabbles are not an unknown phenomenon among security services of virtually all nations, even up to the present day. The 'spat' limited the development of the army's military code breaking unit in the precious months during the opening phases of the war. The main reason for the row seemed to be that the 'Y' stations intercepting enemy signals were gathering a mix of messages of both naval and military interest, so the War Office and the Admiralty agreed to a guarded co-operation. This arrangement broke down as a result of what seems an immature competition between the two services, probably because Room 40 at the Admiralty was making better progress in decoding intercepted messages than the War Office – and flaunted it. The Admiralty was preparing to go it alone in their signals intelligence war.

Wireless telegraphy was a fairly recent invention in 1914, but it had been made a standard installation in all major Royal Navy ships, as well as those of the fast expanding Imperial German Navy. Telegraphy was widely accepted and practiced by sea goers as it could act as a lifeline, helping them to survive the dangers of the sea, but it was just about to become an effective weapon of offence as well.

Germany was loath to risk her navy against a larger British fleet in a major action, so they decided on a strategy of attrition by trying to catch smaller detachments of the British warships in short, sharp actions. Penetration of German naval codes and ciphers by the British using the new technology helped them to counter this strategy. Deception in signals transmissions became a widely used practice by both the British and German navies, often in innovative ways. One aspect of signals security for ships at sea was to disguise the recipient of a radio message, so a common ruse among German transmitters was to direct a message from one coastal station to another rather than to the ship at sea for which it was really intended. The vessel would 'overhear' the message and act upon it; this was the beginning of a wireless-based game of hide and seek played by intercept services to help win or lose the war at sea.

Ewing needed cryptographers urgently so the first place that he looked was within the Royal Navy itself, with staff members of the colleges at Dartmouth and Osborne the first to be interviewed. One candidate was Alastair Denniston, who was teaching German at Osborne. He was offered an appointment which was assumed by all to be a short-term one during the school holidays that he was then enjoying. No one envisaged that the war would last for several years and that cryptology would prove a life-long career for Alastair. He became a major figure in Britain's long chronicle of cryptology and intelligence spanning almost half a century and two world wars. Ewing recruited a disparate team of characters who came from many walks of life, creating a mixed bag of extraordinarily cerebral linguists and naval experts in his team. Room 40 would prove far too small for the fast growing band of decoders, but it did have the advantage of being situated in the Admiralty Building within a few moments' walk of Churchill's and Fisher's offices. Churchill decided that there should be some ground rules for the staff, and his directions, written on a single page of Admiralty notepaper and dated 8 November 1914, is still in Britain's archives. It is headed 'Exclusively Secret' and addressed to COS (Chief of Staff, Admiral Oliver) and D of Educt- (Director of Education, which is a position Sir Alfred Ewing still retained) and read:

An officer of the War Staff, preferably from the ID (Intelligence Division) should be selected to study all the decoded intercepts, not only current but past, to compare them continually with what actually took place in order to penetrate the German mind and movements and make reports. All these intercepts are to be written in a locked book with their decoded, and all other copies are to be collected and burnt. All new messages are to be entered in the book, and the book is only to be handled under the direction of the COS.

The officer selected is for the present to do no other work.

I shall be obliged if Sir Alfred Ewing will associate himself continuously with this work.

The order has the initial WSC in red ink, the date 8/11 and the counter-signature of Admiral Fisher which was an F in green ink.

Paranoia was working fine in the Admiralty.

Codes in the War at Sea

During 1914 cryptologists on all sides were making slow progress in understanding the codes and ciphers of opposing and even other friendly nations. What was badly needed was a stroke of good luck and several of them came at once to Room 40, revealing the working algorithms and keys to codes being used by the Imperial German Navy. They used three main but differing code methods and Room 40 was about to get lucky three times in a row. The first event occurred on the other side of the world in Australian waters where the German steamship *Hobart* was blissfully unaware that war had been declared and was boarded by men of the Royal Australian Navy. She must have been one of the diminishing numbers of ships not to have wireless installed so she had not kept up with world events. The boarding party soon discovered the ship's secret papers hidden by the German captain. Among the documents issued by the Admiralstab (the German equivalent of the British Admiralty) was the HVB code book, or *Handelsverkehsbuch*, which was used to communicate with German merchant ships. A copy was immediately despatched to London for Room 40's attention, so now they could begin to read at least some of Germany's coded transmissions to her merchant ships at sea.

At about the same time in the cold misty waters of the Eastern Baltic, a small force of German cruisers and destroyers were patrolling the waters of the Gulf of Finland close to Russian coastal territory. Dense fog came down and separated the ships of the flotilla and one of them, the *Magdeburg*, was located, attacked and sunk by Russian cruisers in their own territorial waters near their fleet's base in Kronstadt. The Russians later sent down a diver onto the sunken *Magdeburg* and brought up the Admiralstab's most secret SKM (*Signalbuch der Kaiserlichen Marine*) code book bound in heavy lead bindings from the radio booth of the wreck. The book, or more probably books, were dried out and found to be undamaged and quite legible (later on the Imperial German Navy took the precaution of printing code books in ink that was soluble in water). German radio traffic was monitored by the Russian intercept service at Kronsburg and deciphering bureau at Petersburg to prove that the SKM secret code was still in use. This enabled the Imperial Russian Navy to read all the coded radio traffic of the Imperial German Navy. The Russian government generously sent a copy of this code book to London for the attention of Room 40 so that they, too, were able to monitor the transmissions and movements of the German fleet. Both the *Signalbuch der Kaiserlichen Marine* and the *Handelverkbuch* copies are still to be found in archives, the first in the British National Archives at Kew and the second in the Archives and Records Office in Victoria, Australia. An account of this and other events related to the work of Room 40 was the subject of a lecture by Sir Alfred Ewing in 1927 which was reported at length in *The Times* newspaper, much to the disgust of the British Admiralty who wanted to keep the nature of the progress that they had made in cryptology during the war secret.

Room 40's third stroke of luck happened in the North Sea as British warships endeavoured to support British Army units by shelling German troops as they advanced along the Belgian coast. A force of British warships were engaged in this action when several German destroyers were sighted and immediately attacked. All the German destroyers were sunk. Before they went down, the crew put all code books and secret papers into a lead-lined box and threw it overboard. This was standard practice in the case of any vessel being abandoned by its crew to prevent confidential papers being captured. Some time later, a British trawler's nets dragged up a mysterious lead lined box and promptly sent it to the Admiralty. On opening the box the Room 40 staff found, among many secret papers, the *Verkehrsbuch* (VB) code book of the Imperial German Navy. This latest lucky break was so dramatic that it was always referred to as the 'Miraculous Draught of Fishes' by Room 40 staff as they now had a full set of German marine code books. As Sir Alfred remarked in his lecture: 'Thanks to several fortunate accidents the British intercept service had nearly all the German codes in its possession; the rest of the encryptions were solved by analytic means.' These three events together turned Room 40's capability as a code breaker into an equivalent of Bletchley Park after being presented with a working Enigma machine. This intelligence coup enabled Admiralty cryptologists to read almost 2,000 German wireless signals a day transmitted by the Imperial German Navy from that time up to the end of the war.

German radio traffic was carefully monitored by the 'Y' stations in conjunction with D/F plotters identifying the location of the wireless transmissions either on land or at sea. Transmissions recorded by the listening posts plus the sightings of ships or submarines by other vessels or coast watchers were reported and passed to Room 40 for collation and evaluation. The information network and plotting table bore a similarity to that which served the RAF's 'Home Chain' system of radar networks and observer corps reports during the Battle of Britain in 1940. Reports of hostile vessels were forwarded on to Room 40 and plotted, but amazingly the plot did not include the position of Allied ships and their proximity to the U-boats. The position of friendly vessels were only recorded onto the same plotting table later in the war, thus the location of U-boats in relation to vulnerable merchantmen or even warships was not apparent in the intelligence centre until then. Intercepts and other relevant details were recorded and cross-indexed on index cards in a system as part of a huge database of station call-signs and other matters identified in previous intercepts. As the system grew, it was possible to quickly identify most call-signs previously encountered in German wireless traffic or other sources. Thus Room 40's system was able to identify the transmissions of ships in the fleet, flotilla units and shore stations by their already identified call-signs. Operators became so familiar with the German system that they could often predict the call-sign that a particular station would be using as much as a week in advance. Intelligence derived from these intercepts enabled the blockade of the German ports and coast to be maintained with a minimum commitment of British warships as intended movements of German vessels were often known in Room 40 before they left port. Plotting enemy vessels was a very valuable function as the ability to locate the whereabouts and movements of all German submarines at sea on a day-by-day basis would enable convoys to steer away from the predators' torpedoes.

The German naval intercept service that carried out the same function as Room 40 was not comparable in size or effort. It was based in Neumünster, near Kiel, and employed a couple of dozen men and a lieutenant as its senior officer. Room 40, on the other hand, had several hundred men with a naval officer of admiral's rank directing its operations. In spite of this, the German intercept service did have its successes, particularly early in the war, before Room 40 got up to strength. Indeed, the U-9 submarine under the direction of Neumünster torpedoed three British armoured cruisers one after the other off the Dutch coast. The sinking of the *Lusitania* was also directed from Neumünster.

German code breaking was probably on a par with British and even French progress in the field during the second half of the First World War, according to Wilhelm Flicke. Although he did have the benefit of much material that would not be published about British cryptographic achievements until the 1930s. Other European bureaus had their own experience in intercepting and decrypting messages, but Room 40's unique advantage was the development of an information management system within its world class signals intelligence service. The public knew nothing of this until Professor Ewing made his first public pronouncement and then little in detail until Patrick Beesly published the definitive book on the subject, *Room 40: British Naval Intelligence 1914–18*. This revealed the level of achievement that Admiral Hall had made as he built on the code-breaking foundations of Professor Ewing to create a cryptological bureau of a standard that compared with that of Bletchley Park in the Second World War. Indeed, judging by some criteria, it outdistanced 'The Park' in its achievements by the decoding and manipulating of the Zimmermann Telegram which brought America into the war and which arguably ended it. This was probably the greatest intelligence coup of all time; to be revealed later.

The achievements of Room 40 had a number of implications for signals intelligence in the Second World War as well as those of the First World War. Room 40's team created the initial foundations for the intelligence triumphs of Bletchley Park and on which that great intelligence organisation was based. The group that served their apprenticeship in Room 40 under Admiral Hall went on, a decade later, to staff the beginnings of 'The Park' in 1939. Denniston had been a principal figure in the Room 40 team almost from its inception and he managed to cosset his small team of cryptanalysts after the armistice under the name of the Government Code & Cipher School (GC&CS). So for ten years this band of cryptanalysts survived under the suzerainty of a reluctant Foreign Office through all the cost-cutting inter-war years. As war came, they moved into Bletchley Park in the Buckinghamshire countryside, carrying with them an invaluable fund of experience. They transferred a group of experts to tutor the growing band of recruits on the organisation of an intelligence information system and learned how to squeeze every drop of knowledge from German transmissions to feed the complex process of creating perceptive intelligence evaluations. Bletchley Park's true contribution to the war effort was the way that the decoded information was organised and disseminated to those that needed it. The legacy of Room 40 to the development of Bletchley Park is clear, and demonstrates how an intelligence organisation's capability is rarely completely wasted, as we shall see of German intelligence.

The German intercept service also had dedicated personnel whose experience stretched back to the First World War. Indeed, Wilhelm Flicke was one of these. He and his colleagues were trained to receive Morse characters at 120 a minute or better and be familiar with his enemy's wireless style and procedures. A typical fixed interception station in the German Listening Service would have been staffed by between thirty and fifty operators who would have a good knowledge of radio procedures of their enemy and be able to observe and record their transmissions without strain. In particular they had to watch for any quirks or errors in their opponent's transmission, because it was mainly in the errors that the key to breaking a code could be found. All this would be in addition to a number of linguistic experts and other technical and support staff.

Intercept services personnel of every nation needed a special kind of dedication to do the close and demanding work required of them in monitoring and decoding radio signals. The long days could stretch into the small hours spent listening for faint transmissions that sometimes faded into inaudibility with many demands on the patience and maybe even health of the operators. Accuracy in recording Morse code messages consisting of long and meaningless strings of letters and numbers that would mean nothing to the operator took intense concentration. Any wrongly read letter might render the whole coded message unbreakable; in addition, the men or women that were taking down these illegible signals at the rate of at least one a second would not have had the slightest idea of the purpose of their tedious work. In association with the listening post would be D/F operators to identify the direction and compass bearing of the transmitter and calculate the location of the signals station being intercepted. Readings had to be taken very quickly as the enemy transmitter would only be on air sending its message for a short time, and all this had to be reported on detailed forms (probably in triplicate) requiring traffic volume estimates, characteristic traits of the enemy radio operator and many other things. Reports of interceptions were forwarded to an evaluation centre to help build a general radio picture of the forces facing them. As intercept operators began to receive a transmission, a two or three man team using a direction finder device would swing their aerial to detect the direction of the enemy station's transmission signal. D/F operators had not only to be well trained but also adroit at working the loop antenna in response to receiving an enemy call-sign; all this called for exceptional speed of operation. The operator would give the D/F team the enemy station's call-sign to be located and ask the team to take a reading and report the compass bearings to the evaluation centre.

Directionals at Sea

Experiments using D/F applications for the British Army were being made by Captain Charles Round of the Royal Engineers in 1915, who had been seconded to the Marconi Company. Admiral Hall in Room 40 heard about the new technology and sent an urgent request for Captain Round to set up some D/F stations for the Royal Navy. For once the British Army had beaten the Royal Navy to the draw in the SIGINT war. This was a little surprising as relations between Room 40 and the Intercept Bureau at the War Office had deteriorated in the early part of the war to the extent that they rarely talked to each other. With the help of Round, D/F stations, which the Royal Navy called 'directionals', were quickly set up along the east coast at Aberdeen, Birchington, Flamborough Head, Lowestoft, Lerwick and York. Some of these stations also incorporated 'Y' stations to intercept and locate enemy vessel transmissions. Stations were becoming increasingly accurate as the number of D/F stations operating the service increased. They spread from Northern Scotland to Southern England and, as operators became more experienced in understanding the effects of magnetic deviations and other variables, they became more effective. By the end of 1914, Admiral Oliver was able to report to Churchill that they had managed to track the course of a U-boat as it left port and sailed right across the North Sea. This was the beginning of Room 40's ability to plot the position of nearly all U-boats

German Army cryptographers at the *horchstelle* (listening post) at Lauf, near Nurnberg, in the summer of 1940. In the centre in glasses is *Regierungsrat* Wilhelm Flicke, head of this group and responsible for intelligence evaluations. Flicke's successor was *Oberwachtmeister* Suenkel, who is second on the left. They were all probably recovering from the intense signals activity generated during the Battle for France and the British evacuation of Dunkirk. (Roethenbach Military Museum, Germany)

German Abwehr Intelligence Evaluation Centre, Lauf, June 1940, which monitored the despairing radio signals of the French Army in full retreat. The masts seem to have been of a fairly primitive design and were improved later. (Roethenbach Military Museum, Germany)

German officers attending an OKW signals intelligence course in Jüterbog, near Berlin, in October 1944. Left to right: *Oberwachtmeister* Suenkel; behind the tutor, *Hauptmann* Russ, of the Fenast post at Treuenbrietzen; the tutor, *Major* Philiptitsch, also of the Fenast post at Treuenbrietzen; *Major* Wend, commanding the Fenast at Lauf; *Regierungsrat* Wilhelm Flicke in glasses; and, next to Flicke, *Inspecktor* Pokojewski. The Allies had liberated Paris in August and were advancing to the German border. Flicke was tutoring *Nachrichtenhelferinnen* (female operators) at Jüterbog to replace soldiers drafted into fighting units of the German Army. (Roethenbach Military Museum, Germany)

Wilhelm Flicke, working at his desk after the war in 1950 at Lauf listening post. He was financed by the CIA under the command of Gehlen Organisation. The post worked under the cover name of *Bundesstelle für Fernmeldestatistik* (Federal Bureau of Telecommunications Statistics) until the post closed down after Flicke's mysterious death. (Roethenbach Military Museum, Germany)

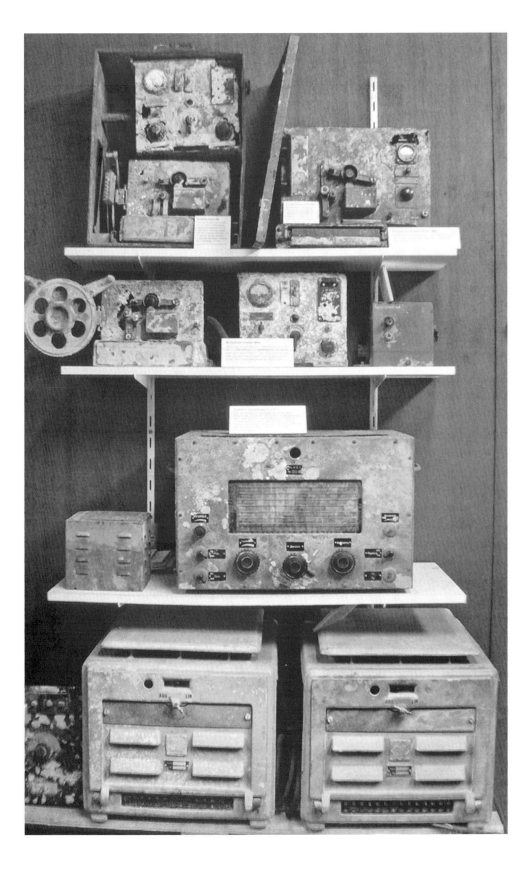

Above: A group of French Resistance fighters in France in 1944. Second on the left is Steve Weiss, who had landed with the American forces in the South of France in Operation Anvil. One of his duties was to stand guard and protect the wireless operator as transmissions were sent to the controllers in Algiers. Later, Steve was one of the few Americans to be awarded the Legion of Honour by the French Government for his services. (Personal collection of Steve Weiss)

Opposite: Radio equipment thrown into Lake Schliersee used at Lauf, as the Americans advanced in 1945. The sets were fished out again in 1950, which accounts for their poor condition. Historic interceptions received on this equipment would have included Colonel Feller's reports to Washington and forwarded to Hitler. Rommel also intercepted them, enabling him to know the detailed plans of the British Army and earning him the title of the 'Desert Fox'. (Roethenbach Military Museum, Germany)

American troops embarking for the Operation Anvil landings in the South of France. On the left, the first man walking up the gangway is Steve Weiss, who joined the French Resistance as a combatant, before returning to face a court martial for being absent from his unit. (Personal collection of Steve Weiss)

The main gate of Spandau Jail in Berlin, where seven Nazi war criminals – Hess, Raeder, Speer, Doenitz, von Shirach, von Neurath and Funk – were serving prison sentences for their war crimes. The prison guard was provided by American, British, French and Russian troops in turn; here British soldiers are shown taking over guard duties from American troops. The author met American TICOM personnel here at this time and began to assist with researching German wartime signals intelligence matters. (Author's collection)

Above: The author's father (left) on the bridge of his ship HMS *Wolf* while acting as escort to a convoy in the Atlantic Western Approaches in 1942. Although he spent much time escorting convoys, he never saw a gun fired in anger due to the guidance of intelligence from Bletchley Park. (Author's collection)

Right: Damage sustained by HMS *Venus*, the author's grandfather's ship, after an engagement with a U-boat in the Irish Sea in September 1915. The vessel was a light cruiser based at Queenstown, near Cork, Ireland. They surprised the submarine on the surface and rammed it, but it did not sink and, instead, limped back to its base in Wilhelmshaven for repairs. (Author's collection)

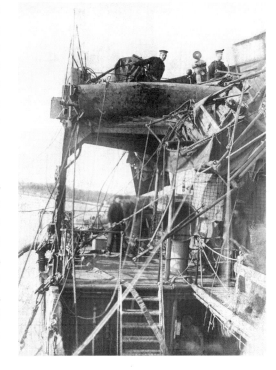

at sea as the service began to improve and U-boat captains still used wireless to talk to each other.

Interception and tracking was only possible when enemy vessels began to transmit but, as the Imperial German Navy were addicted to chattering on their radios, this was not much of a problem. The real difficulty for U-boat hunters was that, once the submarine submerged, there was no effective way to track it underwater as listening and detection devices did not yet exist. Anti-submarine weapons such as depth charges were not available until 1916, in which year they sank just two U-boats, but as both equipment and skills increased, total submarine kills by the end of the war amounted to thirty-eight by depth charge and a further 140 kills partly involving depth charges. Underwater listening technology was only developed to a fairly primitive level by the closing months of the first war; the technique only proved workable in the inter-war years by Professor Bragg. The author's father used to recount how he helped to develop ASDIC (Anti Submarine Detection Investigation Committee) devices as a member of an experimental party from HMS *Excellent* in the 1920s. Further experiments were carried out in HMS *P40* with Sub-Lieutenant, and later Lord, Mountbatten in command. They taped a hydrophone to the end of a broomstick to dip it into the sea in an effort to detect the low-frequency sounds of a submarine's engine. Still, the main anti-submarine weapons in the First World War were the anti-submarine netting, patrols of small craft who were ordered to ram any submarine found on the surface, and minefields, with mines being the most effective weapon.

The combination of directionals and interception of U-boat wireless transmissions for the protection of convoys was going to be the main tool for countering the submarine menace early in the war, but that time had yet to come. Meanwhile, a daily toll of ships sunk while bringing supplies and food to Britain was increasing in numbers. U-boats were slaughtering Britain's merchant fleet, with over 2,000 ships lost during the first years of the war at sea, almost one in four of those that sailed.

As the nascent directional devices began tracking enemy submarines in the North Sea, it became evident that coverage of a wider part of the ocean was going to be necessary as U-boats began travelling further into the Atlantic. Directionals were installed around the Irish coast and their findings were administered by Queenstown in Cork harbour; later they were also installed in the Mediterranean from headquarters in Malta. Prior to the beginning of the war, the thinking among senior naval officers was that submarines were a defensive weapon to protect their bases. The public had the same view, if they ever thought about them at all. Reality came in a blinding flash eight weeks into the war on a calm September morning off the Dutch coast. Three British cruisers, HMS *Abouker*, HMS *Cressy* and HMS *Hogue* were steaming in line ahead, patrolling the Dutch and Belgian coast, when the *Abouker* was torpedoed and immediately began to sink. The *Cressy* came to her aid at once and began picking up survivors when she was torpedoed as well. The *Hogue* then closed on the other two ships as they settled in the water and was also torpedoed; all three ships sank with a total loss of 1,460 men. More men died in this maritime disaster than in the Battle of Trafalgar. The British public had been used to believing that Britannia ruled the waves and were shocked into realising what the U-boat could do to the nation's shipping. The author's great uncle was fond of relating how he survived the sinking of his ship the *Abouker*, then having been picked up by the *Hogue*, only to be sunk again and later picked up by a Dutch fishing vessel. Captain Weddigen of *U-29*, who sank all three ships, went on to be one of Germany's U-boat aces and, with his colleagues, took part in sinking vessels of all kinds and decimating the British merchant fleet. The U-boat fleet would prove to be a fearsome offensive weapon in the war at sea.

Meanwhile, British submarines patrolling the Heligoland Bight scared the German Naval Staff into ordering their *Hochseeflotte* into a defensive role and rarely going to sea. Britain's maritime blockade of German ports and its trade was turning out to be one of Britain's most effective weapons of war. Many Germans were to starve as a result of it, but Britain was also being choked by the U-boat fleet. In the first months of hostilities, U-boats used internationally agreed 'cruiser' rules that required a merchant ship to be stopped, searched and the crew allowed to take to the lifeboats before the ship was sunk. These rules were designed for surface raiders in a more gentlemanly war, but long delays on the surface proved too much of a threat to vulnerable submarine captains. So 'sink on sight' was ordered by the Head of German Naval Staff, Admiral von Holtzendorff, as his staff calculated that sinking 600,000 tons of British shipping a month would cause Britain to be unable to continue the war. The merchant fleets that kept Britain supplied with essential food and the munitions of war were about to be in more danger than ever.

In 1915 British ships still sailed the seas alone, uncontrolled, unarmed and unescorted in much the same way as they did in peacetime. Escorts would be provided for troop ships and some important merchantmen, mainly liners, but the ships that carried British trade, 'the dirty British coaster' of John Masefield's poem, came and went unremarked, uncounted and unprotected. U-boats roamed further and further out and around the coastal waters of the British Isles, the North Sea and even out into the Western Approaches of the Atlantic Ocean. Gradually, the sinking of merchantmen mounted month by month until forty-nine ships, totalling 135,000 tonnes of invaluable ships and their cargoes destined for Britain, were sunk in one month. The merchant ships supplying Britain with essential food and the munitions of war from America's factories and warehouses were being destroyed at an unsustainable rate. The situation looked bleak; something had to be done and the Admiralty was desperate. The solution was staring them in the face and the intercept services were going to play a vital part in its resolution. The convoy system had been proved in the days of sail, but senior naval officers could not bring themselves to believe that the system would work with modern ships. In any case, the Royal Navy preferred to use their warships in aggressively patrolling danger spots and to seek out and destroy the enemy. Meanwhile the U-boats they were seeking were attacking merchant ships that were soft targets and almost severing Britain's vital supply routes. The matter was about to be taken out of the hands of the Admiralty by the politicians.

U-boats and Convoys

The solution was obvious and would give the intercept services a vital part to play in its resolution. The convoy system had been proved in the days of sail during the Napoleonic Wars and even before, in the Tudor period. Senior naval officers were convinced that the convoy system would not work with steam ships and the modern tactics of naval warfare. Warships were to patrol danger spots and aggressively seek out and destroy the enemy, even though results from these actions were proving to be poor. Prime Minister Llyod George visited the Sea Lords to force the issue and insisted that a policy of convoys of merchant ships be immediately instituted. Some senior naval officers had come round to that view already, but by no means all the old guard were easy to persuade. As a result, slow progress was made in fully establishing the policy of escorted convoys of merchant ships before it became standard procedure; indeed, the policy of full convoy protection by the Royal Navy was still not properly established by the end of the war in 1918.

Losses of ships in convoy dropped dramatically, even though at first only home-bound ships to British ports were escorted. Merchant ships still sailing alone were being sunk at an increasing rate compared to their sister ships in convoy. Overall, ships sunk decreased in number and, in turn, the number of U-boats sunk increased as they tried to attack their targets and were themselves attacked by the convoy escorts. However, strength in numbers was not the only advantage that British merchantmen were given. A young U-boat commander, Karl Dönitz, who was to play such a large part in the next war at sea, was captured after his submarine was sunk by British convoy escorts in 1918. He was quoted as saying that the introduction of the convoy system in 1917 robbed the U-boat of its opportunity to become a decisive factor in the war. The oceans suddenly became bare and empty of ships for long periods of time, so that he and his colleagues could see nothing until a huge concourse of ships surrounded by a strong escort of warships of all types would appear over the horizon. The lone U-boat might then be able to sink one or two ships, but then the convoy would steam on, leaving the escorts to search for and attack the submarine. It was this experience that convinced Dönitz that U-boats were not effective as single operators, but needed to hunt together in a concerted method to seek enemy convoys. When a convoy was found, the U-boat group would be able to attack in such numbers that they would overwhelm the escorts. This was the concept that he carried out so successfully in the Second World War in the form of the U-boat wolf packs in the Battle of the Atlantic.

There was another factor to the convoy strategy of which Dönitz was not aware. Room 40 was able to locate the position of virtually all U-boats at sea from an early stage in the war and steer convoys away from the danger. This was probably the main reason why Dönitz found the seas empty. The convoy system made such a tactic simple because every commodore leading a convoy had wireless on board with which to receive warnings and instructions as to which way to steer to avoid a predatory submarine. Communications with other ships in the convoy could be carried out by flags or signal lamps. On the other hand, ships sailing alone were difficult to locate and sometimes did not even have a wireless set on board; this made the protection of ships outside the convoy system pretty well impossible. At last the monumental work of Room 40 was paying off and the evasive routing of streams of merchant ships crossing the Atlantic not only protected them from U-boats, but also from the danger of newly laid mines, both enemy and friendly. The mine was always a weapon to be feared, but sometimes the weapon could be used against the minelayer. Specially designed U-boats regularly laid mines at the entrance of the British naval base of Waterford Harbour, in Southern Ireland, which was a busy port for both merchantmen and naval vessels. Room 40's decoders were able to assess that the Imperial German Navy had cracked the British code that reported which harbour channels had been swept clear of mines. The commandant of the port was asked by Admiral Hall of Room 40 to close Waterford Harbour for a week or so and not to attempt to sweep or clear mines in the harbour channels. Hall then sent a fake wireless report in a code that he knew the Germans had cracked indicating that the channels had been swept and were clear. The Imperial German Navy intercepted the message and despatched the minelaying U-boat *U44* to lay more mines in the channels of the harbour entrance. Instead of laying more mines it struck one of its own unswept ones and was sunk, but some of the crew were saved. The U-boat commander who survived was very bitter under interrogation and complained to all that would listen about the inefficiency of British minesweepers!

Surface Raiders

Germany's Navy did not have the effect on the German people that they were going to deserve in the war at sea. The Royal Navy was well equipped to deal with surface raiders and were numerically superior to the Imperial German Navy. At the outbreak of war in August 1914, the small and well-trained German East Asiatic Squadron, commanded by Vice-Admiral Graf von Spee, faced far superior British naval forces in Hong Kong and Wei-hai-wei. His squadron consisted of the heavy armoured cruisers *Scharnhorst* and *Gneisenau*, and the much smaller cruisers *Dresden, Leipzig, Koenigsberg, Nurnberg* and *Emden*. The admiral left their base at Tsingtao (now Chingtao) in China's Yellow Sea to disappear into the vastness of the Pacific Ocean. He became a thorn in the side of the Royal Navy which was trying to protect the trade routes of the British Empire and the thousands of ships sailing them. Von Spee's manoeuvres started with the despatch of his light cruiser *Emden* south into Australian waters to make an occasional wireless signal to distract and deceive the British ships. The British sought him, communicating with each other by wireless the whole time, while von Spee just listened to them to keep abreast of their movements. Meanwhile, his squadron maintained a discreet radio silence only occasionally punctuated by the little *Emden* transmitting enough signals to convince the hunters that the whole squadron was in Australian waters. The threat had a paralysing effect on British and other Allied maritime trade, and the protection of Australian troop convoys required many of Britain's warships.

The Royal Navy had no idea where von Spee's squadron was lurking so they sent a naval squadron under Admiral Craddock to hunt down the German raiders. Once in the Pacific, the Admiralty sent Craddock's mixed and ageing bag of warships sailing off in all directions in a frantic search for the elusive German squadron. Von Spee had less than a dozen ships under his command but over 100 British and Allied ships were searching for him in the immensity of the Pacific. Craddock had split his forces, taking four of his vessels down to the coast of Chile to the Coronel Islands. He was surprised to encounter a stronger German squadron, but immediately attacked von Spee at the cost of his armoured cruisers HMS *Good Hope* and *Monmouth*, both sunk with all hands. The light cruiser HMS *Glasgow* and armed merchant cruiser HMS *Otranto* fled the field to fight another day. Meanwhile, the Royal Australian Navy had set up a Room 40 type of bureau of their own with the code books captured from the steam ship *Hobart*. There is some uncertainty as to the part played by the Australians in decoding von Spee's wireless transmissions after the battle, but for whatever reason Churchill and Admiral Jackie Fisher decided to send two powerful battle cruisers to seek the German squadron. Von Spee had rounded Cape Horn in the meantime in an attempt to run through the British blockade on the way home to Germany, but was delayed at the Falkland Islands in the South Atlantic.

The battle cruisers that the Admiralty had sent were also to stop off at the Falklands to resupply, so the Royal Navy wreaked its revenge by sinking the *Scharnhorst*, *Gneisenau*, *Nurnberg* and *Leipzig*, along with von Spee and many of his men. The only ship to get away was the light cruiser *Dresden* which many Royal Navy ships tried to hunt down, particularly HMS *Glasgow* and armed merchant cruiser *Orama*. The captured German code book gave an officer of the *Glasgow* the key to a transmission from the *Dresden* saying that she was very short of coal and would be sailing to the island of Juan Fernandez off the Chilean coast. The *Glasgow* sailed with utmost despatch to catch the *Dresden* at anchor and her crew ashore chopping wood for fuel. After a couple of rounds she hauled down her ensign and sent the ship's intelligence officer, Lieutenant Canaris, to play for time so the crew were able to lay charges and sink their ship. Canaris would go on to command the German Army Intelligence and Counter Intelligence agency for Hitler in the next war.

The *Emden* then left Australian waters and sailed to the Cocos Islands to destroy the wireless station there, but the operator managed to send off a wireless message about the approach of an unknown ship. This brought the Royal Australian Navy's cruiser *Sydney* hotfoot to the island to find and sink another of von Spee's raiders, leaving only one from the original squadron. The light cruiser *Koenigsberg* had been located by transmissions decoded by Room 40 and was bottled up in the Rufiji River in German East Africa (now Tanzania) and finally destroyed in July 1915. It had taken just a year to end Germany's *Kreuzerkrieg*, or cruiser war, making the Pacific Ocean safe for Allied merchant ships. Von Spee's little fleet had performed very creditably indeed and its admiral had shown both skill and courage against overwhelming odds. His squadron had tied up over 100 of the Royal Navy's ships and destroyed a number of them. SIGINT had acted for and against each of the combatants in turn, with von Spee maintaining radio silence while using the transmissions of his hunters to avoid them. Finally, the superior capability of Room 40 and Australia's new bureau had directed the ships of the Royal Navy to defeat a tiny German fleet that had certainly punched above its weight.

The North Sea

The strategy of the German South East Asia Squadron in the Pacific had been to harass and hide; this would not be possible in the smaller bounds of the Atlantic and even more difficult in the constraints of the North Sea. The Imperial German Navy knew that if they confronted Britain's Grand Fleet here the odds would be heavily against them. The Royal Navy had a much larger fleet of ships with a tradition of winning their battles, making them a formidable enemy. The German admirals' smaller fleet was better equipped and trained, but against such odds they needed a strategy that would even the playing field. They opted for hit-and-run tactics that would bait the battle cruiser squadron of the British fleet with the hope that they would only meet a single squadron with which to do battle. The German Admiralty's cautious strategy was expected to whittle away an important part of the British Home Fleet's strength. What they had not realised was that, with Room 40's decoders providing advanced warning of the German fleet's intentions to 'come out', the British would negate the effect of the German strategy. The bait that the Germans would lay for the British would be the bombardment of English coastal towns by their fast-moving battle cruisers.

The first attack happened four months after war began, with a scouting force of battle cruisers under Admiral von Hipper shelling Yarmouth on England's East Coast. Room 40 had intercepted the order for a number of vessels to leave their port but their intention was not clear so the town was heavily shelled without any loss to von Hipper. Encouraged by this successful action, Fleet Admiral von Ingenohl decided to give the same treatment to the Yorkshire towns of Scarborough and Hartlepool a month later with cover from a strong squadron of battle cruisers and supported by the Imperial German Navy. The decrypts made by Room 40 could enable Admiral Beatty to intercept the German squadron in an engagement that was set to win or lose an important victory and change the balance of naval power in the North Sea. If von Hipper was successful, the Imperial German Navy would be put on equal terms with the Royal Navy; but if the British destroyed this elite squadron of the Imperial German Navy, then they would have undoubted command of the North Sea for the rest of the war. What the British battle cruisers did not know was that the might of the German fleet was at sea only miles away from their squadron. Their own Grand Fleet was still in harbour in Scapa Flow, and they were in great danger. A confused action took place between the screens of destroyers; this alarmed von Ingenohl as he felt he might be facing the whole of Britain's fleet, so he turned and ran for home. Von Hipper's squadron was left to fend for itself while Beatty was steering towards the sound of gunfire, not realising that he was chasing Germany's main fleet.

Churchill's account of the opening scene cannot be bettered:

On the morning of 16th December at about half past eight I was in my bath, when the door opened and an officer came hurrying in from the War Room with a naval signal which I grasped with dripping hand. 'German battle cruisers bombarding Hartlepool' I jumped out of the bath with exclamations. Sympathy for Hartlepool was mingled with what Mr. George Wyndham once called 'the anodyne of retaliation'. Pulling on my clothes over a damp body, I ran downstairs to the War Room. The 1st Lord had just arrived from his house next door. Oliver (Admiral Sir Henry), who invariably slept in the War Room and hardly ever left it by day, was marking the position on the map. Telegrams from all stations along the coast affected by the attack, and intercepts from our ships in the vicinity speaking to each other, came pouring in two or three a minute. The Admiralty also spread the tidings and kept the fleets and flotillas continuously informed of all we knew … The bombardments of open towns was still new to us at the time. But, after all, what did that matter now? The war map showed the German battle cruisers identified one by one within gunshot off the Yorkshire coast. While 150 miles to the Eastward between them and Germany, cutting mathematically their line of retreat, steamed in the exact position intended, four British battle cruisers and six of the most powerful battleships in the world … only one thing could enable the Germans to escape annihilation at the hands of a superior force … while the great shells crashed into the little houses of Hartlepool and Scarborough, carrying their cruel messages of pain and destruction to unsuspecting English homes, only one anxiety dominated the thoughts of the Admiralty War Room. The word 'visibility' assumed a sinister significance…

Winston Churchill

Mercifully for Beatty, he received a message saying that Hartlepool and Scarborough were being bombarded and so he turned around to try to find the German ships shelling the towns. Von Hipper and Beatty were steaming towards each other at over 40 knots (a knot is 1 nautical mile or 1.1 land miles per hour), but in very poor visibility so as they sighted each other through the fog and then opened fire. Beatty then signalled to a small detachment of his cruisers to disengage, but the whole squadron took it as an order to do so and as they turned away their quarry disappeared into the mist. Thus the best chance that the Imperial German Navy would ever have of dealing the British fleet a devastating blow was lost, but from now on the intercepts of Room 40 would give the Royal Navy a decided advantage in the war at sea.

A month later, in January 1915, a raid by von Hipper to clear fishing trawlers suspected of acting as surveillance units for the British fleet on the Dogger Bank in the North Sea was mounted. Room 40 had deciphered orders for the operation and so Admiral Beatty set a trap with five battle cruisers to engage von Hipper's three vessels. In a confused action, and due to yet more errors in signalling between the British ships, they were unable to engage the enemy properly. Only the hindermost of the German cruisers was caught by Beatty's squadron so the SMS *Blücher* was sunk while the others escaped into the North Sea mists. Again, what the Admiralty and indeed Room 40 did not know at the time was that the German *Hochseeflotte* was at sea and that they had been steaming close to the British battle cruiser squadron. It was the British ships that would have been running into the heavy guns of the German fleet and destruction. Meanwhile, the British battle cruiser squadron encountered the German fleet's destroyer screen and a confused action ensued, leading von Pohl commanding the *Hochseeflotte* to think he had made contact with the British Grand Fleet. Mindful of the instructions coming directly from the German Emperor Wilhelm II himself not to risk his fleet, von Pohl turned and retired, thus missing his greatest opportunity to destroy an elite part of the British fleet.

The Battle of the Dogger Bank had serious repercussions back in the German fleet's base as they realised that this battle of attrition with the British battle fleet was not working. The new Admiral of the German fleet was forbidden to seek battle beyond the Heliogland Bight minefields and the British and German Admiralty were both equally furious at the lost opportunity. In Germany, the blame settled on Fleet Admiral Friedrich von Ingenohl, who was replaced by Admiral von Pohl. Churchill was lambasted by the press for allowing the German bombardment of English towns; for security reasons he was unable to disclose how close the British fleet were to catching von Hipper's cruiser squadron.

The failure to bait British squadrons of battle cruisers into unequal battles left the German Admiralty in need of another strategy, so they turned to unrestricted submarine warfare again. The first submarine campaign had lasted almost a year, but had encountered diplomatic problems while attacking neutral shipping, particularly those of America. The U-boat offensive that would resume in early February would inevitably run into diplomatic problems and worse. Meanwhile, the death of Fleet Admiral von Pohl in February 1916 made way for the more aggressive Admiral von Scheer, who ordered units of his fleet into the North Sea for a renewal of shelling British coastal towns. Lowestoft and Yarmouth were bombarded again and battle cruisers of the Imperial German Navy were planning their next incursion against Sunderland. Foul weather made that attack impossible so von Scheer had to be content with harassing smaller British ships in the Skagerrak (Norwegian waters), supported by the battleships of the Imperial German Navy. Room 40 issued its warning to Admiral Jellicoe that the German fleet was about to leave port and the British Grand Fleet was at sea before the Germans had cast off their mooring ropes.

Jutland

The Royal Navy had dominated the oceans of the world and developed a reputation for successfully tackling its maritime foes for almost 300 years. The British people expected that any confrontation between their Grand Fleet and the German *Hochseeflotte* would inevitably end in another Trafalgar, but there was no Nelson at Jutland on either side. The size of the opposing fleets seemed to confirm that a British victory would happen, although many of the British ships were elderly compared with their opponents and this would become evident in battle. One of the main reasons the British went to war with Germany was their concern about an expanding German fleet which the Germans said they needed to protect their expanding global trade. The confrontation that was about to take place led Churchill to say that Jellicoe was 'the only man who could lose us the war in an afternoon'. There was much at stake.

To von Scheer's great surprise, the two fleets were to meet in the stretch of water between Norway and Denmark called the Skagerrak. Von Scheer was heavily outnumbered, with the odds being twenty-eight British dreadnoughts to sixteen German ones and nine British battle cruisers to five German. The Germans were also outgunned by 270 heavy guns to 200 and, most importantly, the British fleet was appreciably faster than the German one. The order of battle of the two fleets was thus:

THE GRAND FLEET
Admiral Sir John Jellicoe in command

BATTLE FLEET

1st Battle Squadron

Agincourt
Colossus   2 hits
Collingwood
Marlborough   4 hits
Hercules
Neptune
Revenge
St Vincent

2nd Battle Squadron

Ajax
Centurion
Conqueror
Erin
King George V
Monarch
Orion
Thunderer

In addition to these ships: eight armoured cruisers, twelve light cruisers, fifty-one destroyers and a minelayer, of which three armoured cruisers, the *Black Prince*, *Defence* and *Warrior*, were sunk

4th Battle Squadron

Canada
Iron Duke
Royal Oak
Superb
Temeraire
Vanguard

In addition to these ships: five light cruisers (of which four were sunk) and thirty torpedo boats

DER HOCHSEEFLOTTE
Vice Admiral Reinhard Scheer in command

BATTLE FLEET

1st Battle Squadron
Heligoland   1 hit
Nassau
Oldenburg
Ostfriesland   damaged by mine
Posen
Rheinland
Thüringen
Westfalen

2nd Battle Squadron
Six pre-dreadnought ships and one sunk

3rd Battle Squadron
Friedrich der Gross
Grosser Kurfürst   8 hits
Prinzregent Luitpold
Kaiser
Kaiserine   2 hits
König   10 hits
Kronprinz
Markgraf   5 hits

Vice Admiral Franz von Hipper in command of the scouting force of battle cruisers under Admiral Reinhard

Deffinger   21 hits
Lützow   sunk

Benbow

Moltke	5 hits
Bellerophon	
Seydlitz	22 hits
Von der Tann	4 hits

Battle Cruiser Fleet

Vice Admiral Sir David Beatty in command of the 1st Battle Cruiser Squadron under Jellicoe

Lion	13 hits
Princess Royal	9 hits
Queen Mary	sunk
Tiger	15 hits

2nd Battle Cruiser Squadron

Indefatigable	sunk
New Zealand	1 hit

3rd Battle Squadron

Indomitable	
Inflexible	5 hits
Invincible	sunk

5th Battle Cruiser Squadron

Barham	6 hits
Malaya	7 hits
Valiant	
Warspite	15 hits

In addition, the battle cruiser fleet also had fourteen light cruisers, twenty-eight destroyers, of which eight were sunk, and a seaplane carrier.

When senior officers either misinterpreted or asked the wrong questions of their intelligence staff, the cost in battle could be dear. So it was to be at the Battle of Jutland when Captain Thomas Jackson at the Admiralty asked a question and was told the truth and nothing but the truth; unfortunately the answer did not completely address the inadequately framed question. Transmission levels and their decrypts could usually give a number of days' warning to Room 40 that the German fleet was preparing to put to sea. Jackson was generally dismissive of 'all this dammed intelligence nonsense' being plucked out of the air and was not a great fan of the Room 40 operation. The intelligence evaluations he received were often dismissed without comprehending the implications of what he was told, but he had been made aware of the expected departure of the German fleet and that could not be ignored. This irascible senior officer made one of his infrequent visits to Room 40 in 1916 to demand in what location the D/F stations placed the wireless call-sign 'DK' (the call-sign of the German *Hochseeflotte* Commander when in harbour). Room 40 analysts answered his question precisely, 'In the Jade River in Wilhelmshaven, sir'. They all knew that the *Hochseeflotte* was normally berthed there. If Jackson had asked further questions about the wireless security practices of the German fleet, he would have been told that it was the commander's practice to change from his call-sign to another one when his flagship put to sea. German fleet commander's call-signs were then routinely transferred to a shore-based wireless station and another used as he set sail; that way the wireless transmissions plotted by British directionals would not identify his changing location. This routine was followed in the lead up to Jutland, although Room 40 had already deduced from traffic analysis and decrypts that the German fleet was soon putting to sea. Admiral Jellicoe was on alert but assumed that the enemy fleet was still safely in harbour, although he had taken the precaution of taking the British fleet to sea from its Scapa Flow base. He acted on the general alert but was misguided by Jackson's mistaken message that the *Hochseeflotte* were still in harbour; the fleet proceeded to intercept the *Hochseeflotte* at a slow speed to preserve fuel as his ships crossed the North Sea. Two hundred and fifty ships of all sizes, the majority of them British, were to do battle at Jutland. The first contact between the two fleets was made at 2.20 p.m. and not in the early morning light as Jellicoe had hoped. The cannons thundered all through the short daylight hours and into the night as the German fleet turned away under the cover of darkness to seek safety in their home base.

Night actions continued with the British still engaging the rear of the German fleet; confused and bloody they retired. The *Hochseeflotte* sought their base in Wilhelmshaven and were radioed the order to return, giving the course that the fleet should set for home. The decrypt from Room 40 was handed to Jellicoe but he decided to disregard it. He knew that German crews had the advantage of both training and equipment for combat at night. In addition, his faith in cryptographic intelligence had been shaken by Jackson's error and another reporting that the German cruiser *Regensburg* was close by his ship during the battle when he knew that not to be true. (The German navigating officer on that ship had calculated his position as being 10 miles off the actual). A whole string of messages were intercepted by Room 40 during that night, indicating the course that the German fleet intended to make good its escape. The intelligence summaries passed to Jellicoe during the night still failed to convince him of their accuracy and pursuit of the German fleet in the morning light resulted in sinking and damaging a few ships in each fleet; but the British did not get the close engagement Jellicoe desired. He had been given several reports of the German fleet's course before they passed through their own minefields channels and safety. The *Hochseeflotte* made it back to the safety of their berths in the Jade River in time for dinner that evening, having inflicted enough damage on Jellicoe's Grand Fleet to be able to claim a victory.

The British fleet's failure to bring the German *Hochseeflotte* to a conclusive action was largely due to a breakdown in the distribution system of intelligence evaluations and British admirals' lack of faith in their content, coupled with poor communication between British ships (mainly destroyers) during the battle. Room 40 had given several days' early warning of the German fleet's intention to come out and, when it did actually leave harbour, Room 40 had intercepted a great many German transmissions. Those intercepts had revealed enemy movements during most of the critical moments of the battle and the German fleet's course as it retired. Jellicoe had an opportunity to pursue his opponent's ships as they ran for harbour but had chosen not to do so.

The Imperial German Navy's *E-Dienst* bureau, with less than 100 staff and a single junior naval officer in charge, was overwhelmed by the great flood of intercepts coming in during this huge fleet action, even though British W/T security kept their transmissions down to a minimum. Jutland, or as the German Navy called it, the Battle of the Skagerrak, was not the victory that the British people expected of their Grand Fleet. Beginning the battle so late in the day, Jellicoe did not have enough daylight hours to make the most of his battle fleet and a failure to disseminate the superior intelligence gathered by Room 40 properly had robbed him of a victory. So the Germans claimed, with some reason, to be the victors as they had fewer losses than the British. Jellicoe's fleet had lost fourteen ships to Germany's eleven and twice as many men in the Royal Navy died as in Germany's fleet. It is said that it was a tactical victory for the Germans but a strategic one for the British as the German fleet was said never to have challenged the British fleet again. However, that is simply not the case. If anything, the reverse is true as there were a number of forays into the North Sea by the *Hochseeflotte*, such as the raid into Norwegian waters in 1918 to attack a British convoy, but the British Grand Fleet never had the chance of challenging German naval strength again. Only a few sorties were made after Jutland, but Britain still maintained its stranglehold on Germany's economy by maintaining the blockade as one of her main weapons of war. It was thought that Jellicoe could have won or lost Britain the war, though none of these fleet actions seems to have shortened the war at all. It was the winning or losing of the submarine war that could have proved decisive. The public did not know it, but Jutland was not the success it might have been for the British, not because of the failings of the Royal Navy, but due to the mishandling of intelligence. It is SIGINT that would have been the hidden victor had things gone differently and intelligence evaluations acted on properly.

6

The War in the Air

Zeppelins and Gothas

German airships were used to bomb civilian targets extensively during the war; these dirigibles, as the huge airships were called, took their German innovator's name, Count von Zeppelin. They made numerous attacks on the cities of London and Paris with varying levels of military success, but they had considerable psychological effects on the civilian population. London was bombed and other English towns on the east and south east coasts, such as Great Yarmouth, Folkestone, Margate, Ramsgate and Tonbridge, were frequently attacked, as were many others in Northern France. Zeppelins carried out almost fifty bombing raids on southern and eastern England, although many more had to be aborted due to bad weather or engine failure. The Zeppelin's main use, however, was maritime and they were deployed to counter the effects of British ships engaged in the blockade of German ports. Zeppelins made over 1,000 patrols over the North Sea, spotting where the British had laid mines and later to aid the destruction of those mines. A Zeppelin was able to land on the sea next to one of their own minesweepers to show the crew where to sweep and some were even known to hover over an enemy merchant ship to order the crew to their boats and then sink her or ensure that she was taken as a prize to a German port, although there is no record of them doing the same for a warship.

As Zeppelin raids on Britain grew more frequent, Room 40 began gathering information about the bases and movements of these lumbering giants as their wireless signals were as regular as those from the U-boat fleet. They used the HVB code book, of which Room 40 had a copy, and the Germans did not attempt to cipher it in any way. Regular requests for checks on location were made to German direction-finding stations that were intercepted by the 'Y' stations, along with vessels sighting them as they crossed the North Sea. Warnings were often made by Room 40 on details of expected raids, so that a message such as 'an air raid is probable tonight on London' was not unusual. Such information was of use to aircraft of the embryonic Royal Flying Corps, although the poor climbing performance of their machines made Zeppelins very difficult to catch. The BE2 biplanes took some 50 minutes to climb to 10,000ft and, when they did, the ammunition they used was not very effective until incendiary bullets were invented in 1917. Nevertheless, because of these advanced warnings, a number of Zeppelins were shot down or damaged.

The Zeppelin airship developed its technology considerably during the war and served the Imperial German Navy well, although the army version was not so successful. One army dirigible did, however, make an extraordinary attempt in 1917 when German troops fighting in East Africa needed supplies. The *L-59* Zeppelin undertook a trip of over 4,000 miles, which it did in less than four days, to attempt to deliver much-needed food and ammunition to the German forces in Namibia, although when the dirigible finally arrived it could not find their soldiers in the vast tracts of the African bush.

Zeppelins exhibited shortcomings in their function as a bomber so the German raiders started to use the *Luftstreitkrafte* (Imperial German Air Service) Gotha heavy bombers that flew from airfields in Ghent in Belgium. Air Operation *Turkenkreuz* began in 1916 with daylight bombing raids on Folkestone, Sheerness and other towns in Kent. In June 1917, a raid on London caused 162 deaths, of which eighteen were primary school children in Poplar in London's East End. Air Commodore Lionel Charlton's opinion was that it marked the beginning of a new epoch in the history of warfare – how right he was. There were twenty-two raids carried out over South East England in this first 'Battle of Britain' and up to the end of the war 200,000lbs of bombs were dropped for the loss of sixty-one German aircraft. These losses were unsustainable for the German Air Service so they ceased to mount attacks on England and concentrated on the battlefields of France. Zeppelin transmissions were still being intercepted by British stations as they flew over the battlefields of France.

Air Battles in France

Allied aircraft only concerned themselves with reconnaissance at the beginning of the war, so planes of the Royal Flying Corps observed and reported massive

German troop movements across Northern France in 1914. They observed von Kluck's army as it manoeuvred to surround the 100,000 British soldiers in the British Expeditionary Force, causing their dramatic retreat from Mons. A few days later, French aerial reconnaissance also reported German forces moving to the east of Paris and not advancing to attack the city as expected. French troops were rushed to counter this move in Paris taxi cabs and so the German Army was held at the Battle of the Marne.

Pilots needed to land near army units to report their observations and sightings of the enemy to intelligence officers in this fluid war of movement. It was not always possible, however, to find a landing strip near the headquarters of many army units so some pilots tried dropping their messages in packages, but this haphazard method could leave important reports hanging in trees or just disappearing into mud. Wireless telegraphy equipment in aircraft was being experimented with as early as 1915 to transmit observations about troop movements or to act as observers to 'spot' the fall of shot for artillery batteries. Observation was not always reliable, of course. One pilot reported that he had seen troops running away from their positions in panic, but on further inquiry it became clear that the pilot was witnessing a football match. A more reliable form of reporting was needed and so aerial photography was born. Soon, hundreds of miles of the front were being photographed twice a day. Over half a million photographs were taken during the war and the quality of those photos had developed so much that, by the end of the war, it is said that you could discern a muddy foot print from 15,000ft. This was the beginning of another aspect of intelligence gathering that would be used in conjunction with signals intelligence intercepts.

7

The War's End and SIGINT

America Drifts into War

As the war dragged on, America gradually began to drift away from pacifist attitudes and undertook acts that were not entirely neutral. One of the reasons for this shift was the huge munitions orders placed with US manufacturers by the Allies, from which Germany was excluded as the materiel would not have got through the British blockade. Munitions and weapons were supplied through private companies to avoid compromising the United States' neutral status, so the ever-increasing demand began to fill America's factories, employ her people and boost the economy. The American government was well aware of the huge drain on the resources of the Allies and began to fear that they would not be able to meet their bills. Congress decided to raise large war loans to finance the steady stream of shells and guns being supplied to Europe. In addition, British merchant ships, although sometimes armed, were being allowed into American ports, although they would be regarded as warships under international law and interned as such. President Wilson and his government were less than neutral in another key respect as they had to allow the Royal Navy to search and arrest any American ships trying to take goods of any kind into German ports. The blockade was most effective and the German population were almost starving, so the German government resumed unrestricted U-boat war. The United States government had never been truly neutral and the resumption of this kind of U-boat warfare would be another factor in America's decision to declare war against Germany.

The American people still needed a watershed to finally accept the commitment of their country to the fray. There were two events: the first was a maritime disaster caused by either a mistake or a conspiracy; and secondly, the outstanding signals intelligence operation of the war, or indeed in the history of signals intelligence.

The *Lusitania*

The *U-20*'s Kapitänleutnant Walther Schwieger recorded in his ship's log that he had passed Cape Clear on the Southern Irish coast, but did not realise that his submarine had been sighted by a patrol boat. The Admiralty later claimed that notification of the sighting had been passed to the RMS *Lusitania* which was then still 120 miles west of Fastnet Rock. Patrols were despatched to find the U-boat which they assumed was proceeding up St George's Channel towards Liverpool, where there would be many targets for his torpedoes. The neutral ship HMS *Hibernia* had been attacked on 4 May and the torpedo had missed, so it was clear that *U-20* was prepared to attack any ship flying any flag, neutral or otherwise. On 5 May, Schwieger sighted the sailing vessel *The Earl of Latham* and his men boarded her and sank her with explosive charges. The small crew of the sailing ship took to their boats and rowed ashore to inform the Royal Navy base at Queenstown (now Cove in the Irish Republic). The steamship *Cayo Romano*, of South American origin, was attacked but the torpedo missed (Schwieger must have been a bad shot). The *Cayo* docked at the Queenstown naval base and reported the incident, so it had become quite certain that a U-boat was operating on the main shipping route to Liverpool near to Queenstown. At 10.30 p.m. on 6 May, the Royal Navy transmitted a warning from Queenstown to all ships in clear language saying that a U-boat was active off the Southern Irish coast. On that same morning *U-20* sank the SS *Candidate* by gunfire and, in the afternoon, the SS *Centurion* was hit by two torpedoes and sank without loss of life. The danger to the *Lusitania* was now obvious to British naval authorities, as was determined by an inquiry after the event.

Kapitänleutnant Schwieger's log entry for 7 May reads:

1.45 p.m. Good visibility and weather so wait off Queenstown Banks for a target.

2.20 p.m. Sight dead ahead four funnels and two masts of a large steamer steering straight for us and identified as a large passenger steamer.

2.25 p.m. Dive to periscope depth and proceed at high speed on an intercepting course in the hope that the steamer will alter course to starboard along the Irish coast. Steamer altered course to starboard to set a course for Queenstown permitting an approach for a shot. Proceed at high speed until 3 p.m. in order to gain a bearing.

3.10 p.m. Clear bow shot with torpedo set at 3 meters depth with an inclination of 90 degrees with an estimated speed of 22 knots. Torpedo hits starboard side close abaft the bridge, followed by an unusually large explosion with a violent emission of smoke far above the foremost funnel. In addition to the explosion of the torpedo there must have been a second one of boiler or coal dust in the superstructure above the point of impact. The bridge was torn apart, fire broke out and a thick cloud of smoke envelopes the upper bridge. The ship stops at once and quickly takes on a heavy list to starboard and at the same time starting to sink by the bow. She looks as though she will quickly capsize. Much confusion on board as boats are cleared away and some lowered into the water. Apparently considerable panic; several boats fully laden and are hurriedly bow or stern first at once fill with water. Owing to the list fewer boats can be cleared away on the port side than on the starboard one. The ship blows off steam and forward the ship's name *Lusitania* in gold letters is visible. Funnels painted black, no flag on the poop. Her speed was 20 knots.

3.25 p.m. as it appears that the steamer can only remain afloat for a short time longer, dive to 24 meters and proceed out to sea. Also I could not fire a second torpedo into the mass of people saving themselves.

4.15 p.m. Come up to periscope depth and take a look around. In the distance a number of lifeboats; of the *Lusitania* nothing more can be seen. From the wreck the Old Head of Kinsale bears 358 degrees 14 miles. Wreck lies in 90 meters of water. The distance from Queenstown 27 miles at a position 51 degrees 22'6 North and 8 degrees 31'West. The land and lighthouse is very clearly visible.

Most of the 1,700 British and American passengers would probably have crowded on deck watching as the ship made its landfall after a calm crossing. Confusion and panic reigned as the ship was struck by a torpedo. The ship sank far more quickly than the *Titanic* four years before; of the 1,700 passengers aboard, 1,198 men, women and children were to drown in the ensuing panic. Of those, 128 were American and their countrymen were outraged by the attack, bringing the United States another step closer to war. The arguments about the disaster began immediately and each side justified their own version of the drama.

The British took the position that this was the murder of almost 1,200 innocent people and their official inquiry into the sinking came to an understandably anti-German conclusion, and evidence of a second torpedo. The ship was not a part of the Royal Navy; she was an unarmed Royal Mail passenger ship and not conveying troops or armaments of any kind and, therefore, was entitled to 'cruiser rights' applicable under maritime law. The inquiry, chaired by Lord Mersey, stated that 'torpedoing without warning was a further breach of that law' and threw in a few asides about the 'Huns', the use of poison gas and slaughter of Belgian civilians. The language of the report was made even more incandescent by an account that the Germans had struck a medal to celebrate the victory.

Unsurprisingly, the German account differed and stated that they were shocked at the outrage of the American people to an entirely justified attack. The British decision to arm merchant ships with large calibre guns and also tell captains of merchant ships to ram U-boats when on the surface made the 'cruiser' rules untenable. All-out U-boat warfare was a result of the illegal blockade by the British and the *Lusitania* was carrying weapons and munitions which made her a legitimate target for their submarines. They also cited that the *Lusitania* was illustrated and recorded in the authoritative publication *Jane's Fighting Ships* that listed all the world's fighting ships. Lastly, the German Embassy embarrassed the US government by reminding them that they had never warned their nationals that travelling on a British ship was dangerous and that as a neutral they were required to do so. The German Embassy in Washington had, therefore, taken it upon themselves to advertise in American newspapers about the risk of travelling in British ships. However, the case against Germany was made worse in American eyes when the vessels *Arabic* and *Sussex* were both torpedoed and further US nationals were drowned.

However, the US government found the situation awkward because the Germans were indeed correct that they had not warned their own nationals that the trip on a British ship was risky. It was not disclosed until later that the US Customs recorded 1,500 cases of shells and other armaments in the ship's manifest that had been loaded and carried in the hold. The US government suggested that all combatants should return to the use of 'cruiser' rules, which the Germans immediately refused, and that British warships should allow food ships to pass through to their ports to feed the German people who were on the verge of starvation. The British also refused and so negotiations petered out in the heat of war.

All this time, nobody disclosed the fact that Room 40 had actually traced the exact position of *U-20* all the time that it was at sea. It was this and related facts that gave rise to various theories that the British government deliberately arranged to put the *Lusitania* into danger to incite the American people to war. Whatever the true story, America was already moving to war when Germany resumed unrestricted submarine operations on 3 February 1917, causing America to finally break off diplomatic relations with Germany.

Zimmermann and America

As we have seen, intercept and decryption of messages scored a number of successes during the war that shaped battles and even campaigns, but there was only one that decided the course of a whole war. The train of events began in August 1914, a few days after the Germans entered Brussels.

Alexander Czek was a talented radio engineer who came to the attention of the German signals station in Brussels, first by designing an innovative radio set and showing it to a German signals officer and, later, by repairing his heavy duty transmitter set. German officers found that he was not only technically competent with radio equipment, but he was also a good linguist. They employed Alexander and he worked his way up and became trusted in the signals office. During a busy spell when any German staff were off sick, the senior officers readily gave him the job of enciphering messages. The cryptographic bureau was a military one but they also handled diplomatic codes for their Foreign Office and its administration abroad. The *Schlusselbuch*, or code books, of the German Foreign Office, whose code keys were very complex diplomatic codes, needed skill to manipulate them. Czek turned out to be adept at decoding messages as well, so he was used regularly to encrypt messages from Berlin. As he became even more trusted as a member of the bureau, he was increasingly employed in confidential work.

By 1915 Czek came to the notice of British intelligence and the Belgian Liberation Movement, both working clandestinely in the city, and they began to appeal to his sensibilities. They reminded him of his mother's Belgian nationality and how the Germans had attacked Belgium and the atrocities they had committed there. It took a little while to convince him, but he eventually agreed to prepare a copy of the *Schlusselbuch* for the British. This was not easy as he did not have custody of the books and was always supervised, as were all the bureau staff. He finally devised a way to provide 'cribs' by copying the encoded messages afresh, pocketing the rough one and passing them on to his contacts. The work that he was doing did not arouse suspicion, but what did was that he had been seen in the company of Belgian resistance members, so alarm bells began to ring for Czek as well as the Germans. He decided to make a run for it and, after some adventures at the Dutch border, he entered Holland and contacted British intelligence. He was immediately sent to London and, from the information he gave them, the British were able to decipher all German Foreign Office transmissions until the end of the war.

The Imperial German Secretary of State for Foreign Affairs in Berlin was Arthur Zimmerman, who was very anxious about the increasingly anti-German attitude of the United States. By 1917 Germany was concerned at the amount of war materiel that America was sending to the Allies. Zimmermann involved himself in a vigorous campaign against America entering the war and their making materiel and even men available to the Allied war effort, but to little avail. The United States' entry into the war seemed inevitable to him after the sinking of the *Lusitania*, so he decided that he would distract America by making her worried about Mexico. President Carranza of Mexico had friendly relations with Germany so Zimmermann decided to suggest forming an alliance with Mexico against the USA. The first approach to the Mexicans was to be made through the ambassador at the German Embassy in Mexico, von Eckhardt, but there was some difficulty in transmitting a wireless message direct to Mexico City. The German Foreign Affairs Office decided to submit a message to the American Embassy in Berlin, encoded of course, for them to transmit to the German ambassador von Bernstorff in Washington for forward transmission to von Eckhardt. Instructions as to the terms of the approach were sent in a telegram using a diplomatic code which Room 40 could read, and this they did on 17 January 1917. Lieutenant Commander De Gray handed Admiral Hall a partially decoded message, a copy of which can be seen in the British National Archives in Kew, which reads thus:

To the Imperial German Ambassador in Washington
Count Bernstorff

W.158 16th January 1917

For the Imperial German Ambassador in Mexico von Eckhardt

Strictly secret for Your Excellency's personal information and to be handed on to the Imperial Minister in Mexico with ... By a secure route.

We are planning to start unrestricted submarine warfare beginning 1st February. Nevertheless, we are very anxious to keep the United States neutral ... If that is not successful we propose to (Mexico) an alliance on the following basis:
(joint) conduct of the war
(joint) conclusion of the peace ...
Your Excellency should for the present inform the President (of Mexico)

secretly (that we expect) war with the USA (possibly) … (Japan) and at the same time to negotiate between us and Japan … Our submarines … will compel England to peace within a few months. Acknowledge receipt.

Zimmermann.

This decryption was the biggest coup that Room 40 had achieved and would be received by an American government as almost a declaration of war. Even the reluctant President Wilson would not be able to resist going to war with a Germany that was planning to attack his homeland with the aid of Mexico, and maybe even involve Japan as an ally against America. That is, if Hall could convince the president of German intentions with a partially decoded telegram, which they would obviously dispute anyway. Admiral Hall also wanted to use the telegram in a way that would disguise the fact that Room 40 had decrypted it. He might not even need to use the telegram if America were to declare war on Germany as a result of being so incensed at the resumption of submarine warfare. Hall therefore waited for President Wilson's reaction to the new submarine onslaught. Days later it became obvious that Wilson would not let his country go to war over the U-boat war, so Hall decided to act. He showed the telegram first to a staff member in the American Embassy in London and convinced him of its authenticity, and added another one that Room 40 had intercepted which spelt out details of the rewards Mexico could have in the alliance.

Surprisingly enough, Hall was given a free hand by the British Foreign Office to determine the way that the telegram was disclosed to the American government. The contents of this unique piece of signals intelligence was handed to the American ambassador in London by the Foreign Secretary with a request that the contents be kept secret until a cover story was devised. For obvious reasons, Hall did not want the Germans to suspect that the telegram had been intercepted and decoded by the British. The Zimmermann Telegram had been sent as an encoded message routed through the American cable network for various reasons, but this was the height of impudence. Germany was sending an offer to Mexico that they jointly attack America and she was sending the coded text of the offer through the American telegraphic system. When the US government realised this, they seized the encoded text of the cable from the Western Union Office in Washington and, with the advice of Room 40, decoded it on American soil so that the interception would not be traced back to British sources. Attempts were made to discredit the telegram until Zimmermann himself admitted that he had sent it and that the contents were true. The United States of America declared war on Germany on 6 April 1917.

The entry of the United States into the war did not have the immediate effect that it did in the Second World War. America was totally unprepared for war in 1917 and they were not able to field an expeditionary force until almost 1918, but the knowledge that they were coming was a watershed and a wake-up call for Germany.

Meanwhile, the Germans watched the abdication of the Tsar and the struggle between the Communists in Russia with consuming interest.

1918 and the End Game

Even though America's entry into the war did not immediately affect the war on the Western Front for the Allies, it did make an impression on the German High Command. They did not know that the Americans would take so long to mobilise their forces, but they did realise that time was running out for them. The Russians were beginning to show signs of collapse, or at least a lack of will to prosecute the war, and were looking to negotiate a ceasefire. Germany needed to end the war on the Eastern Front quickly so that they could transfer troops to the west for one last decisive push. Over thirty divisions had been moved as hostilities were suspended in the east to reinforce their comrades in the trenches of Northern France. The Allies and, particularly, the French were suffering from a growing shortage of manpower and so, knowing the introduction of the convoy system had enabled the British to survive the U-boat campaign and the war at sea, Hindenburg and Ludendorff decided to seek a decisive victory in France. Accordingly, the *Kaiserschlacht*, or Imperial Battle, was planned with an attack centred on the BEF. They assumed that if the British were defeated, the rest would surely follow. Operation Michael would strike the British Army at St Quentin and the Germans expected to press them back against the sea. On 21 March at 4.40 a.m., as dawn broke, the onslaught on the British positions began.

The crisis was the worst the BEF had faced in the war. General Haig issued an order of the day to his men (still preserved in the National Archives):

> There is no other course open to us but to fight it out. Every position must be held to the last man: there must be no retirement. With our backs to the wall and believing in the justice of our cause each man must fight on to the end.

The situation was held, if not saved, by the arrival of Australian troops and some American troops, although the Americans needed several months' training before being sent to the frontline. The war of movement had returned to the Western

Front once more, although not on the same scale as 1914, and wireless telegraphy was again to play its part. A German Army field cipher named ADFGVX was devised by Colonel Nebel. He chose the letters of the code because they sounded so different to each other in Morse code and, therefore, reduced operator errors. The Germans believed the cipher to be unbreakable, but a French cryptographic officer, Lieutenant Georges Pavin, cracked it in April 1918, a few weeks after the German Spring Offensive. The French concentrated their forces at the point that Ludendorff intended his attack based on a short decoded message, 'rush munitions, even by day if not seen'. Breaking the ADFGVX code helped both French and British generals to counter German moves in the battle. This complex cipher was broken by Pavin, but German operators modified it a few days later and Pavin broke it again within a few days. The emotional stress of working on the code made him ill and he lost 15kg in weight.

The German High Command were about to learn lessons that the Allies had previously learned the hard way in many costly offensives. Telegraphic and field telephone skills used by the Allies in directing artillery with forward observers was a practice perfected by them over constant use in previous years of warfare, allowing them to better protect their troops as they advanced. When German infantry advanced beyond the reach of their artillery cover, casualties would mount astronomically. That was not the only factor bringing the German attack to a gradual halt, but it made a considerable contribution. The effect of railways on rapid troop movements was another; it was easier for a defender to rush reserves to a threatened sector by rail than for an attacker to push forward reinforcements to exploit a breakthrough. The Allies showed initiative in supplying their forward Australian troops at Hamel with food and ammunition by parachute, a relatively new innovation. Indeed, both sides were being more innovative in their methods of combat. The Germans used storm troopers to infiltrate weak spots in enemy lines in a fluid form of attack, but it was a costly tactic. By August, the Germans had run out of steam and the Allies had taken almost 30,000 prisoners. They found that the storm troopers leading the attack had lost the German Army over 150,000 of their best men. The German High Command's last throw of the dice had failed and now it was the turn of the BEF to deliver a killer punch.

The Yanks are Coming

President Wilson of the United States was re-elected in 1916 on a campaign that proclaimed, among other things, that America was too proud to fight. She was making money out of manufacturing the weapons of war for the armies in France, but not preparing to enter the fray herself. As a result, when the Zimmermann Telegrams virtually declared war on the United States, it was a huge surprise. Congress declared war on 6 April 1917 but took some time to send an army formation across the Atlantic. Cryptographers' preparations for the conflict are described in the book *The American Black Chamber* by H.O.Yardley, who had an extraordinary capability to read code. He helped to form the first American Cryptographic bureau in 1918 and had some successes, one of which was catching the German spy Lothar Witzke, who was later executed. The American Expeditionary Force arrived in Northern France at the end of 1917 with little experience or capability in signals intelligence or telephone discipline, which cost them heavy casualties, but in reality they were not in the war long enough to truly learn the lesson that lack of security measures in signals cost in blood. Nevertheless, the appearance of American troops on the frontline was a boost to the Allies' morale and a corresponding blow to the Germans.

Yardley's chamber was later assigned the task of breaking the Japanese code before the naval disarmament conference in 1921, which was a great success by American standards. His organisation came under international pressure after this, so in 1924 its government funding was reduced and only Yardley's commitment to the work of coding kept it alive. In 1929 the new Secretary of State, Henry Stimson, was told of the Black Chamber and with the remark, 'Gentlemen do not read each other's mail', he withdrew the Chamber's funding. Yardley then began making a living writing books and then decoding Japanese ciphers for Chiang Kai-shek in 1940 during the invasion of China. Yardley died in 1958 and, as a pioneer of American cryptology, he was buried in Arlington Cemetery with full military honours.

Brest-Litovsk

After the abdication of Tsar Nicholas II, the Russian provisional government still waged war, albeit with lessening enthusiasm until the Communist Party headed by Lenin took the reins in 1917. Lenin immediately instructed his Foreign Minister Leon Trotsky to arrange a ceasefire and negotiate a peace treaty, which was convened at Brest-Litovsk, near Warsaw, on 3 March 1918. Germany was in a strong position, with Lenin having stated that he needed to finish the war at any price. In the negotiations, Russia lost a great deal of its territory to Germany, much of which was good agricultural land, almost half of its industry and nearly all its coal mines. It was a fearsome price to pay for peace.

General Hoffmann, who led the German negotiating team, had a powerful ace up his sleeve. Telegraphic equipment and a large staff of evaluators, analysts and linguists were quickly assembled to monitor Russian transmissions of all kinds with their government at home. A teletype line for direct access to Moscow was placed at the Russian delegation's disposal by the Germans. What the Russians did not guess was that the Germans had wired in a teleprinter of their own to the system and that their ciphers were being broken by a team of German cryptographers. General Hoffmann was intercepting all the reports and instructions of the Russian negotiators almost as soon as they were able to read them themselves. There were also microphones concealed in the living rooms of the Russian delegation, long before it became standard practice with their own secret service. This gave the Germans a detailed knowledge of the Soviet negotiating team's position, but General Hoffman disclosed his inside knowledge by a slip of the tongue on a couple of occasions during the talks, leading the Russians to change their cipher. This was quickly broken again by the German cryptographic team and, as a result, the Russian negotiators did not stand a chance. Meanwhile, the Ukrainian government (distinct from the Communist government in Moscow) had agreed a separate peace deal to that being negotiated in Brest-Litovsk. This allowed German troops to enter Ukraine's capital city, Kiev, in spite of the treaty with the Communists. Relations between them and the German government were very strained. The Communist government began to form its new Russian Army and made it crystal clear that they would not stand for a separate Ukraine and would use their new army to back up the argument.

The Germans were overjoyed at the settlement, however tenuous, which enabled them to release men and resources from the Eastern Front for the offensive in the west. In some respects, however, it was a disadvantage as German troops from the Russian Front had been infected by Russia's revolutionary ideas which spread to the ranks of their comrades in France. Infections and illness among troops and a hungry civilian population was weakening the resolve of the Germans. Infections, both physical, particularly the great Spanish flu epidemic, and political, spread to other countries, causing revolutionary activity among the populace of Europe for several years. It caused much angst and civil unrest in the newly formed country of Czechoslovakia, as well as Hungary and Bulgaria, but became most widespread in Germany. This was going to resonate into European politics for over a decade and would finally result in the Second World War.

Compiègne and the Armistice

Germany was not the only country in trouble; the Allied countries were almost punch-drunk with the losses that they had sustained and the cost of fighting the war. America wanted to make peace and a separate one if need be, so when the German High Command asked for a ceasefire it came as a welcome bolt out of the blue to the German people, and almost as much of a surprise to the Allies. The German negotiators considered that they might get reasonable terms as they felt their offensive at the beginning of the year left them in a strong position militarily, but they were wrong. The war between the Central Powers and the Allies had to end, so the Armistice was famously signed in a railway carriage in the Forest of Compiègne. The delegation headed by the German politician Matthias Erzberger for the Central Powers no longer believed in the possibility of a German victory and General Foch, representing the Allies, made sure that they met on French soil. When Foch was asked if he had proposals for a ceasefire, he said he had none but if the Germans wanted one his terms were, as an Allied communiqué reaped by the world's press headlined them, as follows:

The vacation of all occupied territories in France and Belgium

The vacation of the west bank of the Rhine

A bridgehead for the Allies into all German territories

The immediate return of all prisoners of war

The surrender of almost all Germany's guns, aircraft and submarines

Surrender of the entire German fleet

Payment of the costs of an Allied army of occupation

Restoration of all damaged property

The handing over of 5,000 railway locomotives, 15,000 wagons and 5,000 trucks

The economic blockade was to remain in place for the time being

The German delegation realised that this was not a ceasefire but a capitulation and the blockade causing severe food shortages and even starvation among German civilians would stay in place until the British had re-established its export links to what it regarded as its own markets. The *Deuxième Bureau* intercept service was, as usual, well prepared to intercept messages about the negotiations at Compiègne, as the Germans had at Brest-Litovsk. All traffic from the German delegation was intercepted and deciphered and when the despatch saying 'Try for milder terms; if not obtainable, sign nevertheless' was deciphered, the Armistice document was immediately signed and the stage for the next war was firmly set. Thus the code breakers' final act in the First World War was to confirm to the Allies that the Germans had no room for negotiation and had, unwillingly, signed the Peace Treaty at Versailles on 28 June 1919. In his memoirs, John Maynard Keynes, the British economist who was a delegate at the conference, called the treaty immoral and incompetent. The South African statesman and later president, Jan Smuts, was quoted in a speech as saying that the peace treaty will eventually lead to a revolution or a new war. The shadowy war of signals would continue.

8

The Inter-war Years

The unexpected ceasefire and Armistice on the Western Front was celebrated with relief by everyone. Germany's soldiers marched home to Berlin through the Brandenburg Gate with bayonets fixed to be acclaimed as unconquered heroes to an uneasy but welcomed peace.

Minor Wars

The uneasy peace did not exist in Eastern Europe, however. There were still savage and bloody minor wars being fought on the Eastern Front for several years after the Armistice was signed. These conflicts were concerned with drawing the borders of the participants of the First World War, only to have them re-drawn again in the Second World War. These smaller wars became proving grounds for SIGINT intercept and decryption skills, and played a significant part, not only in the wars themselves, but in the signals war yet to come. The cessation of hostilities in 1918 on the Western Front brought the fighting in that theatre to an end, but in the east the struggle took on new forms in the Baltic States and along the Polish borders. Large territory claims were made by the Polish government against Germany in 1920 as they created a new Polish Army to contest those claims. In Germany, a volunteer defence force was formed under a National Ministry of Defence in Berlin, staffed mainly by ex-officers of the Kaiser's defeated army, to defend their borders against the requirements of the peace treaty. London and Paris did not think that this was important at the time, as it was on Germany's eastern border rather than the west. Indeed the actions may have been viewed by them as restricting the spread of Communism.

The German radio stations of the garrisons in the region began to operate a signals intercept service to monitor not only Polish troop movements but also those in the Soviet Union where civil war had broken out and even involving radio traffic in Hungary. A 'Volunteer Evaluation Office of the OHL' in Berlin's Friedrichstrasse was formed by Colonel Bushenhagen, recently of the German Army, who had experience of radio interception and deception on the Italian Front. At the same time another ex-army cryptanalyst, Captain Selchow, was setting up 'Bureau C' for the German Foreign Office. In theory the two bureaus were to serve different purposes with Bureau Buschenhagen intercepting military traffic and Bureau C doing the same for diplomatic traffic for the Foreign Office. A duplication of work was bound to arise between the organisations so they became rivals and later bitter enemies. A foundation had been laid by German cryptanalysts to create the great fault of dual-purpose interception organisations that became unwilling to share their intelligence findings or even operational experience. This was to bedevil the effectiveness of German interception services until the end of the Second World War.

Volunteer German border guards established their posts all along Poland's boundary; one army formation was based in the North at Hammerstein and another in the South with its headquarters in Breslau. The newly formed Reichswehr (soon to be the German High Command) commanded both of these mainly volunteer formations, both of which set up their own intercept and evaluation services to use to good effect against the Poles. Signals about the movements and intentions of the Polish Army were intercepted and utilised by the German generals, much as von Hindenburg had done in the Battle of Tannenberg five years earlier. The Polish Army paid dearly for its lack of cryptology skills, but learned the lessons of security in signals much more quickly than the Russians had years before, some of it in a quite unorthodox manner.

Polish Con-Men

In the spring of 1919, at the German Southern Army High Command Headquarters, a major wearing technical insignia on his uniform presented himself at the intercept service offices. He gave his name as Dr Winkler and said he had been ordered by the new Defence Ministry to offer his technical and linguistic abilities to the Southern Army Head of Communications. The good doctor was very sociable and well informed, and soon became easily accepted in the officers' mess. His command of the Polish language enabled him to translate documents and talk about technical matters to members of the intercept service.

He made great play of his contacts in Berlin and was observed telephoning them often while he lived it up in the best hotel in Breslau. This state of affairs lasted for several weeks until some secret documents and maps disappeared and questions began to be asked, so the mysterious Dr Winkler disappeared as well. He left large bills at his hotel and other places and a lot of red faces in the Southern Army. This incident had some far reaching effects; the Poles learned much about the German intercept service, its practices and how to camouflage their transmissions. This was probably the first step the Poles took in creating what became an intercept service comparable with the best in Europe during the inter-war years.

In Germany the two intercept services were maturing with practice by monitoring foreign wireless transmissions and processing the material they intercepted. They established six intercept stations in different parts of Germany during the 1920s. Their skill was further honed by monitoring the frequently held military exercises and manoeuvres of the armies of their neighbours in Italy, France, Austria, Belgium and Holland. Interception of signals in air manoeuvres was a new aspect of signals intelligence and the French Air Force got special attention from the Luftwaffe as they identified the structure and method of France's air warning and reporting system. Aerial spotting techniques of targets and their movements were observed and evaluated, and reports of the German military attachés filed away for future use. The Italians used wireless transmissions in their manoeuvres so much that there was an embarrassment of cryptographic riches. Interception was only limited by the difficulty in mountainous regions of using D/F surveillance systems that were restricted to a short base of triangulation, limited by the Alps. Dutch and Belgian military manoeuvre intercepts were observed more successfully, mainly because there were no mountains in those countries.

During this period, direction-finding and interception stations operated under the guise of marine search and rescue, so each country focused their observations on their national interest. The British were interested in maritime matters, the French in aeronautics, the Germans in army communications.

By 1930 a number of signals intelligence agencies were developing in different ways for differing purposes making the German signals intelligence scene very complex. To simplify the situation it is easier to view it from the post-war perspective, as the Allies discovered for themselves on entering Germany. Their findings offer us an insight into how the German intelligence service readied itself for the war that was to come.

TICOM Analysis

At the end of the war in 1945, the Allies decided to assess the progress of German research and technical progress in nuclear physics, rocketry and jet propulsion. The findings were well documented among academia and made public at the time. An operation that was less well known was the Target Intelligence Committee (TICOM), concerned with the evaluation of German signals intelligence technology and achievements. The project was an Anglo-American effort using staff from Bletchley Park to establish the progress that German cryptographers had made in breaking Allied codes. A prime objective of the investigation was to not let useful findings fall into unauthorised hands, in other words the Russians. The author met some of the TICOM team in Berlin in 1947 when they were winding down their investigations and some of the observations in this book come from that time. Other TICOM operations covered Italy and Japan but there was little that the reports from these countries revealed compared with the rich harvest of cryptographic achievements found in Nazi Germany. The publication of documents in 1986 concerning the agencies mentioned below was hailed as the last great secret of the Second World War, although even today much remains hidden and many documents still classified.

One document that was declassified on 12 December 1954 under provision of the executive order No. 10501 of the State Department in Washington in 1953 was a manuscript entitled *Kriegshiem im Aether* (*War Secrets in the Ether*) by Wilhelm F. Flicke. The author brought it to the attention of the National Security Agency of the American government in Germany just after the war and a copy is now in US State Department archives and another in the Imperial War Museum. Much of the information in this book about the German intelligence service before and during the Second World War comes from this publication and conversations and additional papers the author had with Herr Flicke and others. A number of original documents from those papers are in the author's collection. Some interesting disclosures are concerned with the intelligence organisations founded between the wars, with those investigated by TICOM summarised below.

The Players

Six main German intelligence agencies were targeted by TICOM, although there were more. Allied teams were briefed to find out what they could in 1945 about German co-operation with Japanese cryptologists with whom the Allies were still at war. The results were kept secret for thirty-five years and even today the whole of the documentation has not been released although some is still coming out bit by bit. The interactivity of German agencies was very complex, not to say convoluted in their relationships and procedures. The main ones investigated were: German Intelligence (*Nachrichtendienst*), *Oberkommando der Wehrmacht Chiffrierabteilung* (OKW/Chi), *Chi Stelle, B-Dienst* (*Beobachtungsdienst*), the Pers ZS Signals Bureau and the FA (*Forschungsamt des Reichsluftfahrtministerium*).

German Intelligence

German Intelligence was directed from OKW/Abwehr (Wehrmacht Supreme Command/Counter intelligence) by Admiral Wilhelm Canaris. The Abwehr was formed in 1921 from the OHL volunteer organisation and served as a section of the War Ministry as allowed under the terms of the Treaty of Versailles which restricted the development of German armed forces. The Abwehr and other army units ignored these restrictions even before Hitler renounced the Treaty of Versailles, which led to much subterfuge on the part of army leaders. It was later attached to the newly formed Wehrmacht, as the German armed forces were christened by Hitler, with the function of obtaining and collating intelligence for the army, navy and air force, and also in a counter-intelligence role. Canaris had been appointed by Admiral Erich Raeder in 1933 as Director of the Abwehr, acting as the Supreme Command's intelligence and counter-intelligence. It had five departments, each with a senior army officer in charge, one of whom the author met after the war.

These departments were:

1 Espionage networks   Colonel Jan Piekenbrock
2 Sabotage and 'special duties'   Major General Erwin von Lahousen
3 Counter-espionage and security   Major General Erwin von Bentivegni
4 Administration   General Hans Oster
5 Ausland or overseas operations   Vice Admiral Buerkner

The Abwehr was directly responsible to the Chief of Staff of the Wehrmacht and had over 15,000 people working at the height of the war with signals intelligence departments of its own; in fact Wilhelm Flicke was a part of that armed forces signals communications command although not a soldier himself. Each of the armed forces also had their own intelligence sections that, among other things, initiated the technique of snatch squads advancing with frontline troops to capture secret papers and equipment. The innovation was such a success that the Allies copied their methods later in the war. Abwehr counter-intelligence ran many very successful counter-intelligence operations among secret agents in Western Europe. The development of other international spying and espionage services was never very effective largely due to the powerful opposition of the SS in a savage turf war between the two organisations.

Oberkommando der Wehrmacht Chiffrierabteilung (OKW/Chi)

The Volunteer Evaluation Office of the Army High Command moved from its home in Friedrichstrasse in February 1920 to the German Defence Building in Bendlerstrasse. At the same time it lost its volunteer status, becoming the OKW/Chi (Chi was an abbreviation of Chiffre or cipher). This was the German High Command Decryption Department and nominally the senior signals intelligence agency in the German armed forces. It was in this department of the Abwehr that Wilhelm Flicke spent his war service from 1939 as Auswertungsleiter Regierungsrat, or Director of Analysis, at his base in Lauf. The Allies expected the OKW/Chi to be a centralised agency focused on co-ordinating interception, decoding and evaluating the efforts of the German armed forces. A centre of excellence in signals intelligence was anticipated in a similar way to Bletchley Park. The TICOM team proved them wrong as the OKW/Chi acted as senior intelligence body for German armed forces, but did little to determine policy for the various armed forces, or rather, could not influence the policy of Adolf Hitler who was the leader of the armed forces. When they could not influence the strategic direction of the war, their next main responsibility was to provide a secure communications and coding system at which it failed dismally. They did not know the powerful enemy that they were up against in Bletchley Park, who were not only better at all stages of intelligence evaluations but able to influence the thinking of Winston Churchill, the leader of British armed forces with them. The main role of the OKW/Chi was to act as support for other individual agencies and provide them with high level assistance. As the Wehrmacht was formed, the OKW/Chi was forty strong, rising to 200 by 1939 and, as the war reached its height, about 800 personnel concentrated on strategic and diplomatic matters. The OKW/Chi only had minor successes in breaking codes against Anglo-American communications from 1939, according to TICOM, but even less in the Russian sector. Wilhelm Fenner was a strong senior influence in shaping the development of OKW/Chi from 1922 when he joined the team, but he finished up after the war as a bicycle mechanic.

The German Army Signals Intelligence Service, known as Inspectorate 7/VI, had its roots in the cipher department of the Defence Ministry who had started their own signals intelligence bureau in the 1920s. By 1942 the OKH signals intelligence service had grown and consisted of three sections, 7/VI was the central cryptographic bureau based in Bendlerstrasse in Berlin. They analysed traffic of a number of nations including the USA, Britain, Italy and the Balkans, and thirdly analysis of Russian traffic was carried out by Intercept Control East (HLS Ost) based in East Prussia. In November 1943 the RAF bombed the 7/VI headquarters in Bendlerstrasse, destroying much intelligence information, documents and the ubiquitous index cards. The bureau moved out of Berlin to Juerbourg and began to build up its records and archives of intercepted enemy material again but never made up for the destruction. The 7/VI Inspectorate split off from the OKW/Chi in the 1944 army reorganisation and became known as *Oberkommando ders Heers/General der Nachrlchten Autklaerung* (OKH/GdNA) which served the army. The OKW/Chi was probably the most influential signals intelligence sector of the German intelligence network, although not the most successful. That accolade should be awarded to the German naval *B-Dienst* bureau.

The German field units of the signals intelligence service in the Second World War serving all branches of the army or *Heer* were the *Kommandeurs der Nachrichtenaufklaerung*, known as KONA, which were signals intelligence regiments generally assigned to an army group. Ten KONA units were operating on all fronts in which the Wehrmacht was engaged, with an estimated 12,000 army personnel serving in a signals intelligence capacity at the end of 1943. If just over 1,000 were deducted for administrative work on the group staff that gives just over 1,000 men to a regiment; there were generally two battalions to a regiment although the organisation and chain of command of each regiment varied to meet operational needs. A typical regiment consisted of a Stationary Intercept Company or *Feste Nachrichten Aufklarungsstelle* known as *Feste* and a Signals Intelligence Evaluation Centre, or *Nachrichten Aufklarung Auswertestelle* (NAAS). That centre was served by Long Range Signal Intelligence Companies, or *Nachrichten Fernaufklarung Kompany* (FAK), of which there would probably be two in number, and *Nahaufnahme* or Close Range Signals Intelligence Kompany (NAK). Each of these units worked at army level with the exception of the close range units which would report to a corps commander within the army group.

The Wehrmacht put much effort into interception on the Russian Front, including a complete signals intelligence regiment in Finland presumably against the Red Army in Leningrad. Traffic analysis was used to good effect by the long-range interception stations. An important one was the Field Fixed Listening Post

00313 or *Feste* located at Lauf in Pegnitz, east of Nurnberg. It was one of twenty *Feste* mainly based on German soil for military and diplomatic interception and decryption stations serving the OKH/GdNA. It was operating from the beginning of the war until April 1945 when the staff realised that the war was lost and threw all their radio equipment into the Schliersee Lake. It was later fished out, dried off and can now be seen in a museum at Heinrich-Diehl Strasse 90552 in Rothenbach/Pegnitz which is fitted out as a war museum of Wehrmacht equipment of many kinds including signals intelligence sets (see plate section). These sets are of historic importance as they will have received many important transmissions during the course of the war.

A most interesting find for the TICOM team was the capture of a wide-band receiver for intercepting Soviet high level radio teletype signals which the Germans called 'Russian Fish'. Prisoners of war that the Americans identified as being a part of the team that was developing the apparatus informed TICOM that there was a 'Russian Fish' in a barracks in Rosenheim in Bavaria. The apparatus was found in a number of crates and the German POWs were asked to assemble it, which they did, and the device began intercepting Russian signals traffic immediately. The working 'Russian Fish' machine was shipped to England for further evaluation where Russian signals traffic was received loud and clear, much of which proved to be high-level transmissions of reports and orders between Moscow and Russian armies in the field. The technology was subsequently subsumed into further development of signals interception equipment during the Cold War.

Chi Stelle Air Intelligence

The German air intelligence service began development in 1937 reporting to OKL (*Oberkommando der Luftwaffe*) and grew into a major intelligence force employing 13,000 personnel by the end of the war. Its skill in developing the ELINT technique of reading Allied radio signals from radar, navigation beacons and radio telephone chatter earned it the reputation of producing signals intelligence without the help of cryptologists. After only a few months of war, in December 1939, a *Chi Stelle* station identified a large formation of British Wellington bombers making an attack on Northern Germany, giving location, speed, height and size of the formation to their fighters. Signals intelligence played an important role in air raid defences in Germany as it did in the Battle of Britain a year or two earlier. The Luftwaffe signals service had an administrative centre similar to that of the army, with analysis and administration based in Potsdam. It operated three radio intercept regiments, one in Russia, one in the Mediterranean and another on the western coast, serving the three air fleets of the Luftwaffe. Each had mobile intercept companies, supported by long-range fixed stations, which dealt with anything from weather reports to enemy activity.

Most of *Chi Stelle*'s cryptographic effort in Russia was based in the field, reading low-grade traffic of the Soviet Air Force and yielding some tactical advantage. Higher grade Soviet traffic using one time pads and SIGABA could not be read. The western *Chi Stelle* stations provided the main air raid warning system as Anglo-American raids got bigger and heavier, although the Luftwaffe information handling system needed to act very fast. Air intelligence had to intercept and decode signals in hours to be of use to air defences, whereas the army could sometimes afford to take days to clarify an intercept. *Chi Stelle* also had to cope with high technology ELINT signals such as radar and navigation aids. Nevertheless, they could measure the strength and bomber types of the attacking force. Allied bombers could often be identified as they took off in Britain by their voice transmissions and then tracked to their target. Sadly for the German cities about to be bombed, there were not enough anti-aircraft guns or fighters to combat the incoming raids. This is a clear illustration of the principle that intelligence is of little use without the force to take advantage of the knowledge.

B-Dienst (Beobachtungsdienst)

The signals intercept service for the German Navy reported to OKM (*Oberkommando der Marine*). SIGINT in the Imperial German Navy (disbanded in 1918) developed during the First World War in a similar way to when Hindenburg 'overheard' the Russian transmissions at the Battle of Tannenberg. The war was being fought in the trenches by the army and field telephones had taken over much of the wireless transmission. German radio operators had time to spare and found they were listening to Dover patrol and the Royal Navy, much of it being in clear text. This was reported to the Imperial German Navy, who saw an opportunity to set up monitoring stations on the Belgian coast to intercept signals from British cruiser squadrons, convoy sailings and shipping across the English Channel. In spite of the intelligence advantages, *B-Dienst* only maintained a small body of cryptographers, supervised by a junior officer in their listening station – nowhere near the effort put into the interception of German naval signals traffic by the British.

As the First World War ended, the Germans surrendered their fleet to the British, with no need to maintain a naval signals intelligence unit. Naval staff realised a signals intelligence bureau would be of great use to them in rebuilding their navy anew. They recruited veterans who had served *B-Dienst* during the war and formed a nucleus of less than a dozen cryptographers to recreate the service. Even this small team was difficult to maintain during the lean years until Hitler came to power in 1933 and ordered a reorganisation of the German Navy only a year later. In the inter-war period the German Navy established two chains of radio stations, one covering the North Sea and the other the Baltic coast. They were controlled from Wilhelmshaven and equipped with the most modern intercept systems, which would later prove critical in the war at sea. The intercept system was backed up by a talented team of cryptanalysts, including academics and businessmen, as well as naval personnel in a similar way to Room 40. The reputation of the bureau was enhanced by an English-speaking cryptanalyst, Wilhelm Tranow, who cracked the Royal Navy code in 1935. As war started, *B-Dienst* grew rapidly until there were over 6,000 working on the Allied naval codes. Tranow's team continued to break Royal Navy Administrative Code in 1940 and the British Merchant Navy BAMS code in 1942, allowing U-boats to track convoys until 1943, at which point the British changed their ciphers. *B-Dienst* was based in Berlin up to 1943 when their headquarters was bombed with the loss of much documentation. (This seems to have happened to virtually every important German intelligence agency during that year.) The bureau moved several times to avoid air raids and the advancing Russians, but finished up at the end of the war in Flensburg. One objective of the TICOM teams was to find a device being developed for the U-boat fleet called the 'Kurier'. This apparatus would condense a submarine's coded wireless signal to just a few microseconds in length. The problem this would have posed for the Allies was that they could not locate such a short-lived transmission using their existing direction-finding methods. This posed a very serious threat to Allied naval intelligence as the location of enemy vessels depended on a fix on a D/F device. The sea trials of the Kurier took place in 1943 and, although the technology was found to require further development, it became operational as the war ended. Like many examples of German advanced technology, the Kurier system reappeared when the Soviet Navy developed a burst encoding system of their own for submarine communications in the Cold War. *B-Dienst* had a record unmatched in German intelligence and was largely responsible for the great success of the U-boat fleet up to about 1943 – a worthy opponent of Bletchley Park in the signals intelligence war at sea.

Pers ZS

The Pers ZS Signals Bureau was the German Foreign Office's bureau with only a small team of cryptographers, but, according to the TICOM team investigating it, they evinced an extraordinary degree of competence. The reason for this was that there was a consistency in its development that was not to be found in any other German interception service. It was the oldest of the German signals intelligence bureau, beginning its operation well before 1914.

Some of the personnel had been recruited in the First World War and were retained after 1918, with some still serving with the bureau at the end of the Second World War. This gave Per ZS a great deal of accumulated experience and expertise. Their task was to intercept and decode the codes and ciphers of the diplomatic messaging systems of over fifty countries. They were able to read the transmissions of most major embassies, including Britain, the United States, France, Italy, China and Japan. They had a team of cryptanalysts with an ability to crack diplomatic codes that was outstanding, particularly those of their allies Italy and Japan. German diplomats, however, were not as effective at encoding messages, as the Zimmermann Telegram debacle revealed. A Soviet diplomatic code book that the Finns had captured was given to the Germans and later taken by TICOM.

The Forschungsamt des Reichsluftfahrtministerium (FA)

The FA was the intelligence arm of the Nazi Party and originated in the weeks after the Nazis assumed power in 1933. Gottfried Schapper, a dissatisfied employee of the Defence Ministry's cipher bureau, went to Reichsmarshall Hermann Goering with the idea of a new centralised civilian signals intelligence agency. Goering sensed an opportunity to boost his own power in the party and in the country with his personal intelligence agency. Goering set up the 'Research Bureau of the German Air Ministry' using his Ministry's resources to do it, although he kept the organisation quite separate from the Air Ministry and used it to function as the Nazi Party's (and his personal) SIGINT agency. The FA was given several objectives: to maintain surveillance of the German Army and in particular its senior officers; to watch leaders of the National Socialist Party as to how well their loyalty to the party could be relied upon; to maintain a watch on the Catholic Church in Germany and its communications with the Vatican, particularly the financial transactions between them; and to target people 'of interest' to the party who had been in political life in Germany including trade unionists and other labour organisers.

All means of communication in Germany were monitored by the FA which surveyed every type of radio and wire communications, particularly those with a connection beyond the German borders. Newspapers, magazines and published items of interest along with the interception of letters and conversations by means of secret microphones were reported to Goering. All government employees above a certain level had their conversations intercepted along with party workers, even those in high places. The author visited the headquarters of the FA called the Haus am Knie in Berlin's Charlottenburg district just after the war; even though it was in a ruinous state after fierce fighting the remains of the electronic communications equipment indicated a very sophisticated operation.

The FA began their work in co-operation with the Cipher Bureau of the Ministry of Defence and Bureau C of the Foreign Office but they soon had their differences and parted. The Gestapo, on the other hand, took an increasing interest in this formidable instrument of domestic intelligence and repression. An early operation of the FA was to monitor the Brownshirts and particularly Ernst Roehm, who was watched constantly, creating a great file of material that was finally laid before Hitler. The intercepts showed all sorts of political sins, real and imagined, so the 'blood purge' of Roehm and his Brownshirts, known as the Night of the Long Knives, was ordered by Hitler in 1934 to tighten the Nazis' grip on the German nation. The FA also played a major part in the German *Anschluss*, or takeover of Austria, by penetrating the entire Austrian communications system. Those organisations or individuals creating difficulties for the German occupation were identified by radio transmissions and were dealt with in advance by friends in the Austrian Army, police or even the Austrian Nazi Party to make the takeover by the German Army a smooth one.

Goering also found the FA useful in eliminating his opponents and a great boost to his standing in the Nazi Party. The FA was a principal weapon in creating the authoritarian state that Germany became. As an instrument of repression, it attracted the envy of Himmler and his Gestapo so it created a strong and growing rivalry between him and Goering which escalated and would become evident later in the war and during the attempt on Hitler's life.

Intelligence has many strands and that of signals and spies often interweave, so it was with Thilo Koulen who was related to a senior officer on the Army General Staff and worked for the Ministry of Defence Cipher Bureau. Koulen was dismissed from there but soon found a place in the FA at a level that enabled him to visit signals intercept stations and discuss cipher algorithms and keys with friends who trusted him in the military intelligence community. He became a visitor to many of them as he chatted about his relation on the General Staff and their coding secrets. This went on for a couple of years until France was defeated and the Germans entered Paris; there they found a freight train full of French government documents. These were shipped to Germany and evaluated, particularly those from the *Deuxième Bureau* who listed the various agents that they had in Germany before and during the war. Thilo had systematically betrayed the whole cryptographic system of the German Army to the French, enabling them to read virtually all transmissions of a military nature for a number of years leading up to the war. Also the Abwehr had 'planted' a spy on the French from Thilo's bureau to provide them with secret documents that were 'doctored' to be of no use. Thilo told the

French how they had been doctored and so provided them with another source of secret information. French intercept services were given a unique opportunity to penetrate German thoughts and intentions and would have undoubtedly shared them with their allies in Bletchley Park. Thilo was arrested and tried for treason. During his trial the judges were appalled at what had leaked out but the prisoner hanged himself in his cell before the verdict was reached.

The TICOM team had been able to identify virtually all the German intercepting agencies operating during the war but said they were surprised to come across the FA network, about which they knew nothing. Either the team's right hand did not know what the left was doing with regard to access to German encryption keys or they were telling fibs, which is not entirely unknown in the intelligence world.

Whatever the case, 'Goering's Research Bureau', as it was nicknamed by senior Nazi officials, actively sought attitudes of dissention among Nazi Party members or the German populace but had little activity in military matters although it did evaluate enemy weapons and machinery. It had a diplomatic signals section which was able to read Chamberlain's messages to London when he was in Munich during the Crisis in 1938; also the Swiss Interbank Code had been penetrated, allowing German bankers to understand the financial strategies of Swiss bankers on whom Germany's international currency availability and dealings depended. The primary function was monitoring press and commercial traffic, but the FA's main source of intelligence came from telephone taps, of which the agency had over 1,000 at one time. The TICOM were tipped off to the existence of the FA by a member of the Pers ZS agency and immediately visited the airfield where the FA was last operating. Most of the FA documents had been burned, although there was enough left to understand the purpose of the bureau. If those destroyed documents had been rescued, the scandals that might have been uncovered would have been a treasure trove for historians with a salacious taste.

Among the agencies not investigated by TICOM was the *Deutsche Reichspost* (DRP). Although it was a civil and not a military organization, the Forshungstelle, or DRP's telephone intercept research bureau, managed to unscramble telephone conversations between Churchill and Roosevelt. No security breach occurred, however, as they both seem to have talked in codes and code words.

Cipher Machines

A number of machines to encode and decode signals were being developed during the inter-war years. The reason for automating the cryptographic process was the growing volume of traffic that an army (or navy or air force) would generate in the course of its operations. The machines also speeded up the coding, preparation and clarification of messages and hopefully made for better accuracy in their transmission. These machines carried a weakness that sometimes made them insecure, as anyone who knows anything of the story of the Enigma machine is well aware. A machine invariably has a repetitive pattern in its make-up and the trick is to find that pattern, which is not easy. A plethora of encoding devices were designed either before the Second World War or in its early stages so there seemed to be almost as many versions of cipher machines active during the war as there were different codes and ciphers. Nevertheless coding machines were a novelty among the military in 1939. Most of them had a rotor mechanism working on similar mechanical principles to the Enigma machine, in which the content of the message was scrambled into an almost infinite number of variations. To complicate things even more, most of the rotor machines had an electro-mechanical aspect which varied the rotor mechanisms with the complication of variable plug-in wiring circuits. Such machine-made complexities could only be solved by another machine and the BOMBE was designed to do this by Alan Turing at Bletchley Park. The machines listed below are grouped by their country of origin; Germany mainly used four machines and Japan depended on one, but each one of them had their code broken by the Allies.

Enigma

There was almost an industry in the manufacturing of coding machines in the inter-war period by all the countries most involved in cryptographics. The first of these machines is the most well known but is probably more universal than is generally realised.

Enigma was not a single machine but a family of electro-mechanical rotor-driven machines with variations for different uses of which 100,000 were made. The Germans used them widely at the lowest level of command throughout Wehrmacht operations. It was a difficult machine to operate as it needed three people to code and another three at the other end to decode a message. A version was sold to the Italian Navy and another to the Swiss Army. Even the Japanese had their own model. The German Condor Legion intercept unit was equipped with a version of Enigma in the Spanish Civil War in the proving ground to prepare the German military and air force for the war to come. The design of a rotor cipher machine was an inspiration to designers of the Typex, SIGABA, the Swiss NEMA and even after the war the Russian Fialka machine. However, the machine had a coding weakness that was broken by the effort and genius of Biuro Szyfrow, a Pole, with some help by Hans-Thilo Schmidt, a German spy who stole a code book. The Polish contribution was not only of importance for its pioneering code-breaking work but, just as vitally, it convinced others that the Enigma code could be broken. Enigma was not a high-level machine but carried many routine reports of personnel postings, ammunition states and tactical battlefield reports. At the war's end a great many Enigma machines were available in Germany. The author was offered several but the price asked of 100 cigarettes was too expensive at that time, in addition to which its significance had not dawned on me. A Mauser 9mm pistol seemed a better bargain at the time but that was later confiscated by British customs.

Lorenz SZ40

This was much more important as an instrument with only about thirty of them made according to TICOM, who later called it Hitler's Blackberry. It relayed his personal thoughts to his most senior commanders and, therefore, the transmissions were of the highest value to Allied intelligence. It was a rotor stream cipher machine used by the German Army in the Second World War from 1942 to 1943 exclusively for High Command messages from Berlin to commander level of armies in Europe. It was designed to protect land-line transmissions, although it could also be used on wireless transmissions. Transmissions from Lorenz were first received in the 'Y' station based in Denmark Hill in London's Camberwell by the Metropolitan Police. They reported 'New Music' to Bletchley Park but did not have the resources to record it properly so a new 'Y' station was constructed in Knockholt in Kent to concentrate on the new transmissions. The breaking of the Lorenz code was done with the use of the Colossus machine at Bletchley Park and was broken when it was realised that the design was partially based on the mechanism of a teleprinter. The relative importance of the Lorenz machine may be judged by the fact that, according to TICOM, each was to be placed with the highest Wehrmacht commanders. A weakness found in its coding by Bletchley Park was that it had a repeat pattern in every forty-second letter of its output. The breaking of the Lorenz code was much more important to Allied intelligence in the closing years of the war than Enigma as the Lorenz was used at a much higher level in the command structure. As the war drew to a close, one of the Allies' important objectives was to capture a Lorenz machine before the Russians got to them.

The Geheimschreiber, or Secret Writer

This machine was developed by Siemens for the German High Command in the Second World War and was used entirely to transmit over land lines initially. It was known as T 54 and was more secure than the Lorenz machine, far more efficient than Enigma and took the invention of the world's first computer to read its output. An operator sat at the machine and typed in his message and the machine encoded it by itself, thus avoiding any human aspect by using online encipherment and then transmitted by land line or radio. It was sent at a rate of sixty words a minute to the machine at the other end which would decipher it automatically. To read it required the genius of the British Post Office who built a Colossus machine with 1,500 radio valves whose working efficiency was threatened by a leaking radiator. If you want the whole story you will find a reconstructed model of Colossus at Bletchley Park and a real *Geheimschreiber*, captured in the desert of North Africa, on show in the Imperial War Museum.

T 43 Schlusselfernschreibmaschine

This machine was also made by Siemens and was known as 'Sagefish'. The German Army designed it to protect teleprinter signals and it was one of their main cipher machines to be used beside Enigma. The code was unbroken during the war but TICOM is thought to have captured a machine. The question is, did the Soviets find one as well? After the war an improved version called T 52 was used by the French and Dutch armed services, but both British and Swedish code breakers were able to read most of their traffic.

Purple

This was the name given by the Americans to the Japanese electromechanical stepping-switch cipher machine used by them during the war because the intercept forms were in purple folders. American decryptions and intelligence that came from breaking the Purple machine code were called Magic by Allied intelligence officers. The Japanese ambassador to Berlin used one reporting back to Tokyo on German preparations on the beach defences in the run-up to the D-Day landings. This confirmed that their deceptions to mislead the Germans as to planned landings were working. The Imperial Japanese Navy did not co-operate with the Japanese Army in cipher machine development because of the competitive spirit between them. The Japanese Army warned the Navy of security risks and weak points in the encoding machines but their advice was ignored. The Germans also warned the Imperial Japanese Navy about radio security but they refused to accept that there was a breach in security until a debate in Congress after the war revealed the secret. The Imperial Japanese Army went on to develop their own cipher machine, based on the principles of the Enigma, called the 92-shiki injiki.

The Allies seemed to do better in machine cipher design although how much better it is difficult to tell with all the claims that are made.

SIGABA/ECM

This was an electromechanical rotor-driven cipher machine developed surprisingly as a joint United States Army and Navy project in the 1930s and used throughout the Second World War. It was surprising because the army and navy were bitter rivals and each developed their own separate cryptographic system and shared very little of their intelligence findings with each other. SIGABA's development was an exception to that rule but, having achieved a little co-operation, the army called their version of the machine SIGABA while the navy called theirs ECM (Electro Cipher Machine) and both had their own facilities each with a print facility of its own. It is said that the SIGABA/ECM code was never broken and as a result it was used by the military well into the 1950s. The German cryptographers called it The American Big Machine and did not differentiate between the army or navy model.

M-209

This was an American machine with rotors and a plug in mechanism to encipher each character of a message and the US military purchased 140,000 of these sets in 1943. The Germans broke the code and, in spite of the Americans knowing that, the sets were still used on the battlefield because they were light and were quickly and easily managed by a single operator. The M-209 became the workhorse of the American infantry and it just shows that security is not everything, at least on a tactical level.

Combined Cipher Machine

The Allies experienced a considerable problem with the British Typex and the American ECM (or was it SIGABA?) machines not being compatible. Design work on this new machine, which made the M-209 and Typex work in a compatible way, cost $6 million and became operational by 1943. It is thought that the Combined Cipher Machine was secure and its code never broken, although some weaknesses were identified by Allied cryptanalysts and swiftly corrected.

Typex

This was a British coding machine developed as a replacement for their code-book system which was slow and awkward and even insecure. The RAF developed their own model Typex machine that became operational in 1937. It was entitled 'RAF Enigma with type X attachments' and was an adaptation of the commercial model of the Enigma but with seven rotors (the German Enigma machine had four). A Typex machine was captured by the Germans but without the rotors in the Battle of France in 1940. Efforts were made to crack it by the *B-Dienst* but were soon abandoned as it was said that the German four-rotor Enigma was undecipherable so the British seven rotor machine would be even more so. The Typex machine was seen to have advantages over the German Enigma as it only needed one man to work it while the German Enigma needed a team of two or even three. Typex did not suffer from operator copying errors as the text was printed on paper tape and could be linked to a teleprinter for printing out, whereas the Enigma machine transmissions had to be written down, enciphered and then transmitted by Morse. Typex machines had their message typed onto tape as dictated which was enciphered and transmitted automatically at the rate of twenty words a minute. The same advantages were true in the reception of a transmission so the efficiency of the machine was distinct, particularly in stressful conditions in action. The machine was used by all the British services and it was estimated that 12,000 were made during the war.

The IFF Code

Also known as the Identification Friend or Foe Code challenge, this was the way British ground control was able to tell a friendly aircraft from an enemy. It was not a messaging system, rather an encrypted electronic signal requiring the pilot of a friendly aircraft to make a valid response by just pressing a button. If an aircraft neglected to do this the controller's radio message would be 'Make your canary sing' but if the pilot still did not respond he would be told to press the tit or he would be shot down. The IFF code was the precursor to the transponder which is fitted into all modern aircraft.

The NEMA

Also known as the Neue Machine, this machine was built by the Swiss when they discovered that both the Germans and the Allies were reading their messages on the version of the Enigma machine which they had purchased from Germany. It was one of the Enigma variations (sounds like Elgar) and inherited many of its generic encryption weaknesses. Intercepting Swiss messages was more important than you might expect as many of the transmissions were about high finance supporting the war economies, particularly Germany's. Gold was the main financial resource which the Nazis used, with bullion taken from the national exchequers of conquered countries and gold stripped from Jews in the death camps.

Soviet Machine Codes

The author has been unable to identify any coding machine used by the Soviets during the war similar to those of other nations described above. The only one that has been found is the Fialka Machine or M125 which is a complex rotor-driven machine whose design seems to have been based on the NEMA Machine described above. It was in use during the Cold War after 1945 but almost certainly does not seem to have seen service during the Second World War.

Non-Belligerents

During the Second World War, Spain and Japan (while it was still neutral) were benevolent in their neutrality to Germany early in the war and provided fertile ground for intelligence gathering. The United States and to a lesser extent Sweden, on the other hand, offered favourable support to Britain but were still an area of intense activity for intelligence agents of both sides. Spanish territory was used by the Axis Powers to set up radio monitoring and ship-reporting facilities and they even helped to supply U-boats on the mainland and Atlantic islands, in spite of repeated Allied protests. On the other hand, control of Spanish and to a lesser extent Portuguese cable and telephone utilities were used by the British to intercept German voice and data intelligence. When gleaned, it was generally used to disseminate misleading information to create friction between Germany and her allies.

A Swede supplied the British naval attaché in Stockholm with German codes and in 1941 the same attaché received a valuable early warning about the German battleship *Bismarck* putting to sea. Neutrals sought to maintain as even-handed an approach as possible to both sets of belligerents so covert links were maintained between all intelligence services of belligerent countries. Intelligence services on both sides sought to derive advantage from such opportunities such as the reading of Turkish ciphers which were used unmercifully to deceive their opponents.

Swedish intelligence tapped into German teletype lines connected to Norway and passing through Swedish territory, which enabled them to read German communications at a time when they were worried about a German invasion. The Swedish government was also concerned about being embroiled in the war after Hitler's invasion of Russia. Sweden did began to co-operate with the Abwehr when Soviet submarines began to sink Swedish ships later in the war. This co-operation led to the Abwehr and the *Sicherheitsdienst* supplying the

Swedes with listening devices to be planted in the homes and offices of Allied diplomats. At the same time the Swedish government allowed Polish Military Intelligence in London to courier funds to their Home Army in occupied Poland until it was discovered by the Gestapo. Many Swedish agents were arrested as a result by German security agencies at the same time as Russian attacks were being made against Swedish shipping. As a result, collaboration was extended to the Axis powers against Russia. It was not easy being a neutral in a world war. Finland signed an armistice with Russia, ceasing hostilities, which led to the Finnish encryption service not being able to operate in their own capital Helsinki. Permission was granted by the Swedish government to move an intercept and decryption service onto Swedish territory for monitoring Russian signals traffic as long as the Finns supplied them with any intercepts that they had made.

Intelligence services on all sides were very active and the prize was not just intelligence but sometimes trade in strategic materials. The British were very short of steel ball bearings as they did not have the capability to manufacture them but Sweden did. Several high powered British motor launches made the perilous journey through the Baltic to Stockholm to pick up large consignments of ball bearings and bring them back to Britain to ensure the continuing production of tanks and guns. The Swedes also had expertise in the development of coding machines of a similar nature to the Enigma; one of these was designed by an inventor Boris Hagelin in the 1930s. He approached the Germans about it but they said that they were not interested. He then tried the American government and as a result the American SIGABA machine was developed whose coding method was never knowingly broken. Germany's most important trading partner however was not Sweden but Switzerland.

Before the war started in 1939, the brilliant Nazi Minister of Economics Dr Hjalmar Schacht warned Hitler that Germany was going bankrupt and that she would not have any foreign currency left by early 1940. The Swiss banking conduit to the foreign currency market needed to be kept open as the Germans wanted foreign currency. Big international investors in German industry had to be kept happy and many of these were American; Allen Dulles, soon to be head of the CIA and the brother of John Foster Dulles, settled in Berne to look after the investments of Americans who owned large slices of German industry. American investments in IG Faben Chemicals and Opel, owned by General Motors, among many others, were producing tanks and guns and even Volkswagens as the Wehrmacht needed to be kept moving. The investments were producing big profits as well, but the trouble was that cash to buy raw materials with which to make them was running out. That is, until the first of the 'Melmer Shipments' which were delivered in suitcases by a Captain

Melmer of the SS, containing gold in many forms, including rings, necklaces and ornaments of all kinds. These items were melted down by Degussa, a company specialising in refining precious metal. The first twenty bullion bars of 999.9 purity were marked with the stamp of the Reichsbank and were delivered on 20 November 1942. Many more cases marked *Konzentrationslager*, or concentration camp, were to be delivered by Melmer and returned to him in bullion form to be sold for foreign currency to buy Germany's weapons of war. No record of the bullion bars was kept, but today the banks of the world hold many gold bars from these 'Melmer Shipments', including, it is said, the Bank of England.

The author was told a story by a Belgian journalist about what happened on 10 May 1940 on a quiet Sunday in Brussels. That morning the German Army crossed the border and invaded Belgium. The directors of the Belgian National Bank were roused from their beds to attend an emergency meeting in the bank's building in the Grand Place in Brussels. What should they do? The German Army would be there in hours they thought and they were the trustees of Belgium's stock of gold bullion. Looking out on to the cobbled square one of them noticed that an excavation was being made in the square to lay some pipe work which the men would complete and fill in when they started work again on Monday. Very early on that fresh May morning the group of elderly directors, some of them still in their pyjamas, turned out to dig the hole a bit deeper and then they carried the gold reserves of Belgium, bar by heavy bar, to be buried under the pipe work. It was not dug up again until the Allies rolled into Brussels in 1944/45 and liberated the city. A journalist might say this story was too good to be true, but it illustrates the fact that the Germans had the gold reserves of the countries they invaded available to them as well as Melmer's suitcases. It was because of the importance of the Swiss connection that the Nazis maintained such a close watch on Swiss signals traffic.

9

The Second World War – The Beginning

Czechoslovakia and Poland

Preparations for war against these two countries were made with great thoroughness by Hitler as he came to power in the 1930s. His immediate target was to integrate the German-speaking minorities in both countries into the Third Reich, but his longer term objective was, of course, the domination of Europe. The *Volksbund für das Deutschtum im Ausland*, or People's League for German Culture, was tasked with recruiting people of consequence in the German-speaking regions of both countries to aid the German takeover and pave the way for Hitler's troops to invade. Eavesdropping on the telephone was a major part of monitoring the Czech people and their institutions, such as the police and army. Cables that touched German, Austrian or even Hungarian territory were tapped; there was close co-operation between Germany and the largely pro-German Hungarian government in monitoring Czechoslovakia. This followed much the same pattern as that established in the Austrian *Anschluss*, where there was a mixture of internal subversion and external threat, and a close watch was therefore kept on the Czech armed forces. By the end of the 1930s very little occurred in the administration of the country and its army that was not known to German intelligence.

After many negotiations carried out in bad faith by Hitler, the country was gradually taken over and on 15 March 1939 his troops marched into Czechoslovakia. Britain's Prime Minister Neville Chamberlain concluded that appeasement had come to an end. Guarantees of military support were made to Poland by Britain, but the move was years too late to save the peace. The Nazi regime was not the first to make preparations for war against Poland, however. Plans for an invasion of Poland had been made by the German Wehrmacht in 1923, if not before, and Hitler simply implemented the plan. Preparations were extensive, including monitoring and infiltrating all aspects of Poland's armed forces by wireless interception and through spies and Polish deserters. The organisation of the Polish Air Force, which used radio freely, was clearly understood due to its intercepted messages; not only was the nature of its organisation known, but details of each plane by type and condition was recorded by the German listening posts. When Poland was attacked she did not stand a chance and the German assault was complete in a few weeks. As their country was being overrun, Polish cipher staff proved their importance to Europe's future as they fled with their secret codes and work through Romania to France to re-establish their organisation there. Then, as France fell, they escaped to England to help set up a cryptographic intelligence undertaking that would influence the conduct of the war and prove worthy of the highest praise. Poland's contribution to fighting many aspects of the war was more than the Allies could have expected.

The Phoney War

In the winter of 1939 the war was not phoney for the peoples of Poland, the Baltic States, the Finns, the Danes and the Norwegians as they all came under savage attack from the Wehrmacht or the Russians. However, the war's first months were an intense preparation for the inhabitants of Germany and Britain as they left their civilian life, collected their uniforms and reported to their barracks, ships or began to dig air raid shelters. The women had to cope with food rationing and some with evacuating their children. The author remembers those busy first weeks of war in his home town of Portsmouth as navy reservists in ill-fitting uniforms were mobilised to drill on vast parade grounds as the war rolled on in slow motion. Men came together (the recruitment of women had not begun yet) to fight a war that seemed as yet unreal. The first shot, or rather torpedo, was fired at the ocean liner SS *Athenia*, on course for America with well over 1,000 passengers on board. The *U-30*, commanded by Oberleutnant Julius Lemp, sank her on the evening of the first day of war on 3 September 1939 and she was the first ship of twenty-six to be sunk in that month. This was the aspect of the war about which Churchill worried the most, as to lose the Battle of the Atlantic would mean cutting Britain's lifeline and seeing his country starve.

In October *U-47* crept into what the Royal Navy thought was a safe haven for their fleet and torpedoed the battleship *Royal Oak* with the loss of 800 of her crew. The loss of one of the Royal Navy's capital ships was a devastating blow to her home town of Portsmouth as its crew were fathers and brothers in the community. The author recalls the stunned grief of Portsmouth; everyone knew at least one of the 800 families that had lost a son or father. This was only the first of many such sinkings. HMS *Hood* was sunk with 1,500 and only three survivors in the northern Atlantic. A month after the sinking of *Royal Oak* came retribution. A signal from Commodore Harwood's cruiser squadron in the South Atlantic to the Admiralty in London said, 'Am engaging German Battleship'. Although the fight was not as bloody as many other sea battles in the war, it left the *Graf Spee* a smoking wreck in the port of Montevideo. The score then was one all, with one capital ship sunk on either side.

Norway

The next drama was only partly naval when British and French troops, including marines from Portsmouth, were sent to land in neutral Norway's northern port of Narvik. The stated purpose of the landing was to send an expeditionary force to help the Finnish Army against the Russians. This wildly improbable plan had a more practical hidden agenda to it – to deny Germany the supplies of iron ore coming through the port from Sweden. In addition, they wanted to make sure that the port was not to be used by the Kriegsmarine; Admiral Raeder wanted to use Narvik as a base from which his submarines would threaten Britain's Atlantic supply lines. The Germans invaded Norway for several reasons, one of which was the *Altmark*, which had been a supply ship for the *Graf Spee*, now sunk in Montevideo harbour. They had transferred prisoners before the battleship was scuttled and the vessel was now trying to evade the British blockade by sheltering in Jøssingfjord, well inside Norwegian territorial waters. Captain Vian of the destroyer *Cossack* decided to rescue the prisoners and so entered the fjord with a naval boarding party, pistols in hands, with the cry of 'The Navy's Here' to capture the *Altmark* and release 300 Allied seamen. Hitler was furious so he ordered the High Command to prepare plans to invade Norway and, because Denmark was in the way, invade that country as well. British aerial reconnaissance photography showed the port of Kiel crowded with shipping and airfields filled with aircraft in preparation for something – but what? Both sides decided to move at the same time with the British laying mines in Norwegian waters while the Germans planned a full-scale invasion of the Norwegian coast.

The Norwegian invasion, codenamed Operation Weserübung, was an improvised operation for the Germans who threw together a small signals intelligence unit drawn from personnel of various other units. Facing them was an even less prepared Norwegian Army whose intercepted signals were almost all in clear language, although most messages proved of little tactical value to the Germans. Messages between Norway and London were of more interest to them, although they could not be decrypted, but frequency analysis gave strong clues as to the main disembarkation ports of British troops. Badly disguised code names and call-signs gave further clues to the chain of command and composition of British movements. The signals intelligence contest was won by the Wehrmacht more by accident than design, with their intercept company performing modestly well. The lack of co-operation between British and Norwegian operators created encryption weaknesses in their operations to the German advantage. The opening stages of the operation were covered from the town of Husum by a German intercept station on the Danish border. The German government were more concerned at what would be the attitude and reaction of the Swedes to the invasion. Swedish reactions were monitored by reading their transmissions from a listening post in Als in Denmark as that country now was occupied. It was not until ten days after the operation started (due to lack of shipping space) that the German intercept company finally moved closer to the scene of operations in Oslo.

B-Dienst naval operations were more effective and they needed to be as Germany was using almost all its surface fleet in the operation to their great cost. British intelligence failed to foresee the invasion and so was unable to catch German troop-ship convoys at sea but two sea battles had a profound effect on German war plans. There were naval actions that seriously weakened the German surface fleet, particularly at Narvik where they lost one heavy cruiser, two light cruisers and ten destroyers, plus a pocket battleship damaged and out of action for over a year. The cost to the British was also high because the Admiralty ignored a warning based on frequency analysis of German signals indicating that German warships were about to put to sea. The omission caused the aircraft carrier HMS *Glorious* to be caught on the way back to England in the North Sea by the German heavy cruiser *Scharnhorst*. The aircraft carrier was sunk with the loss of over 800 crew members, and the Admiralty never ignored an indication from traffic analysis again. Although the Norwegian Campaign did not stem the supply of iron ore to Germany, it did so weaken the German surface fleet that the Kriegsmarine was no longer able to intervene at Dunkirk or be a serious factor in attempts to invade Britain later in the year.

The evacuation of British and French forces from Norway after its surrender on 1 May 1940 left some agents and saboteurs behind to fight a resistance battle, and it also cost the Germans 300,000 soldiers needed to guard the country for most of the war. In early 1940 about fifteen agents were transmitting to London, as counted by German listening stations, but by 1941 transmissions appeared all over Norway. They participated in an organisation called Scorpion, reporting German shipping movements along the coast. The number of agents grew until in 1941 there were over 100 and soon no ship movement along Norway's long coast line would go unreported. One of the coast watchers reported sighting the *Bismarck* off the coast, creating a chain of events leading to its being intercepted and sunk by the Royal Navy. Any German vessel at sea off the northern coast of Norway would expect to be attacked by Russian aircraft or submarines after being spotted by British or Russian agents. Determining the location of those resistance agents by direction-finding methods was difficult due to the mountainous terrain of the country. German Storch light aircraft were used to spot agents and their transmitters but were never very successful and soon another factor in resistance appeared. Russian agents, controlled from Murmansk, were being put ashore by submarine and operating low powered radio sets at 80 to 100m making them difficult for German monitors to pick up. In spite of this, the Russian network was penetrated by German counter-intelligence and it was largely taken over until Murmansk became suspicious. Events in Norway, however, were just a sideshow as the war's focus shifted south.

The Battle for France and Dunkirk

May was an eventful month in 1940 as the Norwegians surrendered on the 1st and, ten days later, Hitler's Panzers rolled into the Low Countries. Meanwhile, the large French Army and the much smaller British Army had waited patiently behind well prepared fortifications for almost nine months to be attacked by the German Army, which took time to recover from the invasion of Poland. But, once ready, the Blitzkrieg, or lightning war, struck hammer blows to both the French and British forces. The attack's preparation included making ready the German Army intercept networks as French, Dutch and Belgian orders of battle behind their frontier defences were clearly discernible to the Abwehr from signals intercepts. German mobile intercept units were reassigned to the Western Front as the Polish campaign came to an end with orders to advance with the Panzers to maintain the best reception of enemy signal traffic. These intercept services spent their time familiarising themselves with enemy formations and their locations. A French command cipher was broken, revealing organisational flaws in French divisional dispositions to German interceptors in some detail. The German High

Command was able to identify concentration areas of both British and French armies from field radio stations practising their Morse code skills, although not their wireless security skills. The boundaries of British Army corps and division were only occasionally identified but they formed a much clearer picture of French forces behind the massive fortifications of the Maginot Line.

The British Expeditionary Force (BEF) had ten divisions under the command of General Lord Gort, but the Germans were able to identify only five semi-motorised and one armoured division by signals intercepts, although they suspected that there were several more. The British order of battle, chain of command and number of troops was not discerned by Germany's listening posts but the Dutch or the Belgian armies were a different matter. The Dutch were careless with their radio security and they also had plain clothes 'visitors' from over the border always asking questions of a penetrating nature. The Low Countries were of special interest to the Wehrmacht as that was the route intended to turn the flank of the Maginot Line fortifications. They did it in the First World War with the invasion of Belgium and were about to do it again. The French had not fortified the Belgian frontier for political reasons, as they did not want to antagonise a friendly nation. Using Belgium as a trap door into France was such an obvious strategy that General Gamelin, the French Commander in Chief, designated his Seventh Army to march into Belgium as soon as the Germans crossed the border. The BEF would march alongside them and hold the line at the River Dyle to the east of Brussels. The Dutch strategy was to withdraw to 'Fortress Holland' while the Belgians would hold the line at the Albert Canal. 10 May was the curtain raiser for the relatively short but savage Battle of France as the Luftwaffe destroyed the Dutch and Belgian air forces on the ground and rail connections into France to disrupt communications. The German Army moved west and signals interception by Wilhelm Flicke's intercept station at Lauf confirmed the Anglo-French Army were moving up to the River Dyle to dig in.

German paratroopers landed to capture Dutch bridges and landing fields for transport planes. After four days of fighting, the Dutch received an ultimatum; if they did not surrender, Rotterdam would be destroyed by the Luftwaffe. Hours before the ultimatum expired the threatened bombing took place, killing and injuring over 30,000 civilians. Lauf's interception of Dutch military radio traffic gradually ceased about 15 May as her tiny army was swept away by the German Panzers.

The attack on Belgium was simultaneous with that on the Dutch. The powerful fortress of Eben-Emael, with a garrison of 2,000 men, was taken by less than 100 German paratroopers. The Belgian Army retreated behind the Dyle River line. General Rommel crossed the River Meuse on 13 May, followed by General

Guderian's tanks further down the river near Sedan the next day to drive their Panzer columns deep into France. The French and British marching in Belgium were in danger of being trapped as Panzers breaking through at Sedan began to cut across their rear. The German plan to cut the Allied forces in two succeeded brilliantly as they sped at 40 miles a day through France, stunning the French Army and French government who were paralysed with fright.

German intercepts were beginning to pick up the command networks of the French and British armies in Belgium. Desperate signals pleading for food and ammunition were heard as the force faced the threat of envelopment. The BEF retired into one of three large concentrations of Allied troops in the Pas-de-Calais area. Two were French, centred in the region of Lille and Roubaix and the third was mainly the BEF to the east of Dunkirk. Every troop concentration was using a great many transmitters in a very limited area; as a result German interception stations could no longer differentiate between them or identify them by D/F methods. A direct radio contact was intercepted between British Army Headquarters and their War Office in London on 22 May while the French Army Group commander was in lengthy communication with the French High Command. In spite of intense efforts, German cryptanalysts could not break the code but another code breaking event happened on that day and it would begin a profound effect on the future of signals intelligence in the war. Bletchley Park was able to read the GAF Red Key (German Air Force general key) only occasionally until 22 May 1940, when the code was then broken and remained readable to cryptanalysts at the Park with few exceptions for the rest of the war. It was the beginning of Ultra, although this tender plant would take over a year to grow into the signals intelligence weapon that would influence the war, beginning as the Battle of El Alamein was fought in the desert. Still, 22 May 1940 was the first step in a long and successful journey of detection and deception of the German High Command by the reading of its most secret signals.

Secret intelligence would have been of little use to the Allies at Dunkirk as the battle was already lost; decryption of a signal from the British 5th Division indicating preparations for evacuation of British troops was no surprise to the Germans. The Dunkirk evacuation, started on 27 May as an ever-shrinking perimeter held by British, Belgian and French troops, was savagely attacked by German Panzers. The British held the west and south while the Belgian Army manned the northern boundary of this embattled enclave. The pressure was relieved by Hitler himself as he sent a signal in clear text to his generals timed at 11.42 on 24 May 1940 ordering his tanks to stop on the line of the Aa canal. Meanwhile, Boulogne to the south was held by the Brigade of Guards, and Calais by a small scratch contingent commanded by Brigadier Claude Nicholson, about to be reinforced by battalions of the Rifle Brigade: the 60th Rifles and the Queen Victoria Rifles, with elements of the Tank Corps. Most of this courageous bunch of men was hurriedly shipped into Calais Harbour from Dover on the *Maid of Orleans* and other cross-Channel ferries. Boulogne was soon evacuated on Churchill's orders which he later said he regretted. Now all that stood between advancing Panzers from the south and the troops on the beaches of Dunkirk was an ill-equipped British garrison in an old French fort. Thrown together at short notice and ordered to hold the Citadel until relieved, the British force awaited the onslaught. At Lauf Wilhelm Flicke's team were receiving 1,000 intercepts a day and one of them was a signal meant for the British commander at Calais.

Inside the Citadel's massive walls surrounding an inner courtyard that is now a football pitch stood a wireless truck of the Royal Corps of Signals. Lieutenant Austin Evitts was receiving a message from the Secretary of State for War in London, whose copy is still in the National Archives:

1415 hours 25th May 1940
To: Brigadier Nicholson Commander Calais

Defence of Calais to the utmost is of vital importance to our country and the BEF and as showing our continued co-operation with France STOP
The eyes of the Empire are upon the defence of Calais and we are confident you and your gallant regiments will perform an exploit worthy of any in the annals of British history
Time out 1415 hours

This was as close as a Whitehall Warrior (an officer in the War Office who has never seen action but directs other in battle) would ever get to saying that you have got to fight to the last man and the last round of ammunition. Just to make sure the message had gone home, that evening Churchill himself sent a further message to Nicholson:

1946 hours 25th May 1940

To: Brigadier Nicholson Commander Calais
Every hour you continue to exist is of the greatest help to the BEF. Government has therefore decided that you must continue to fight. Have every admiration for your splendid stand. Evacuation will not (repeat not) take place and craft required for above purpose are to return to Dover.
Time out 1946 hours

Churchill

Participants in the battle were unaware that the German Interception Station at Lauf was intercepting their messages, although the information would not have made much difference to the tactics of either the Panzers or the Citadel garrison. There was to be little room for manoeuvre in the confrontation yet to come. Whitehall received disturbing news from Major Alexander now commanding the 1st Rifle Brigade after the death of its colonel.

Citadel a shambles STOP Brigadier's fate unknown STOP Rifle Brigades casualties unknown STOP Being heavily shelled and flanked but attempting counter attack STOP Am attempting contact with 60th fighting in the town STOP Are you sending ships

In the Citadel a telephone rang in a deep cellar with a call from Sir Edmund Ironside, the Chief of the Imperial General Staff in London, and was answered by Lieutenant Hugo Ironside, who was a relative. Sir Edmund confirmed that the garrison would not be evacuated and later a British destroyer battled its way into Calais Harbour with a confirmation of the order to hold their position indefinitely. By mid-afternoon on 26 May the defenders of Calais were running low on ammunition, food and water and under increasing attack from German armour and storm troopers. Finally at 4 p.m. the Citadel was overrun. Today the debate goes on as to how important the siege of Calais was to the Dunkirk evacuation. General Guderian thought it mattered little as his Panzers advanced on Dunkirk's beaches, but Churchill thought that the three days the defence of Calais bought for the little ships was critical. Hitler's stop order for his tanks was essential to relieving pressure on the BEF, and in getting away they had left almost everything except their code secrets in the wreckage. The British Army must have been more meticulous about destroying decrypted messages than is generally assumed in accounts of the Retreat from Dunkirk, as the Germans never found a hint of the fact that their Enigma machine code was being read. The German Air Force General Key that had so recently been broken, and a copy of which must surely have been sent to BEF Headquarters, was not found. The precious Ultra secret was still safe.

The evacuation succeeded by the skin of its teeth and the fall of France took place soon after. The frightful slaughter of the Battle of Flanders came to an end. Throughout Germany the flags fluttered and the bells rang out as the German people celebrated what they imagined was the end of the war. Dunkirk was one of several victories for the forces of the Third Reich; in their minds the British could not possibly resist the might of the German forces now and must capitulate. Germany was convinced that complete victory was theirs and the Wehrmacht even began to demobilise some of the 'heroes' that had won the war. Yet the weeks passed and nothing happened. German intercept services reported that the British messages they received showed not of surrender but of foolish resistance. The Reich Chancellery was sorely puzzled and had to tell their people that preparations for an invasion were taking place. German newspapers began to print stories about the effect that bombing would have on London if they continued to resist.

When Hitler's conviction that Britain would surrender faded, he issued Directive 16 for the invasion of South-East England, the plans for which were to be submitted within thirty days. The Abwehr estimated that the British Army could muster eleven infantry divisions, another eight that were understrength and one armoured division (most of the tanks had been left in France). The movement of British troops to the battle area would have to be by train as most of their transport had been left on the beaches. The Chief of the Wehrmacht, General Jodl, planned landings of thirty divisions, stretching from Portsmouth around the coast to Ramsgate and the Thames Estuary. In the huge list of materiel that he needed, he estimated they would carry 2,600 bicycles and many horses as most of the transport would be horse-drawn. The German High Command were planning a major seaborne assault as though it was a river crossing.

The Kriegsmarine knew better and wanted a much narrower bridgehead with landings of three divisions between Folkestone and Beachy Head. The larger landing envisioned by the German Army would need 400 ships and attendant ancillary vessels such as tugs. Even so, they did not have the landing craft to use in the shallow water around the English coast.

The first stage would comprise of a single wave of boats of all shapes, sizes and speeds with landings in shallow water that would require some soldiers to paddle ashore on rafts. The fastest speed that a convoy could do was 3 knots, which meant that the shortest crossing between Calais and Dover would take eight hours. The first wave of the assault would be without reinforcements for about sixteen hours while the ships returned to France to pick up more troops and make the slow journey back to the landing beaches. The Wehrmacht and Kriegsmarine could only agree on one thing: the whole enterprise would only be possible if the Luftwaffe had complete domination in the air.

The Luftwaffe prepared to destroy the RAF and then batter Britain into submission by night bombing with the help of ELINT.

Britain's Air War

ELINT's new weapon was radar in a fairly primitive form, and its importance in both the war in the air and at sea started with the Battle of Britain. Hitler ordered the Luftwaffe to take on the British fighters in the air and on the ground. The Luftwaffe had won air superiority in Northern France and now baited the RAF by using their Stukas to attack shipping in the Channel. Using the new radar technology, Fighter Command was aware that above the Stukas there lurked a squadron of Messerschmitts and would not allow their precious fighters to engage in such an unequal fight. The RAF were going to fight over their own territory and on their own terms, so the air battle shifted to the airfields in South-East England and it was here that the fighter planes of the RAF had to be brought to battle. The RAF's home-ground advantage was their use of radar as an ELINT weapon, harnessed to a surprisingly effective information network that in turn connected directly to the RAF's Fighter Command. The author vividly remembers that glorious summer in 1940 as a schoolboy, as the newspapers reported the number of German aircraft shot down like cricket scores in a great game, though the score was not just one-sided. The Battle of Britain raged from July to September and the Luftwaffe came close to defeating the RAF, until an accident forced a change in the Luftwaffe's strategy. Both sides had tacitly agreed the capital city of the other should be 'out of bounds' but one dark night a German bomber lost its way. It jettisoned its bombs at random without realising they would fall on London. The site where those fateful bombs fell is recorded in stone in London's Barbican. Retaliation by the British was swift with a raid on Berlin, which infuriated Hitler. He ordered the Luftwaffe who, by now had the RAF on the ropes, to switch from attacking the fighters and their airfields in England to 'Blitzing' Britain's cities. As autumn turned to winter the British realised that winter storms would batter an invasion fleet more than their armed forces would; an invasion attempt was no longer possible. This was confirmed by intercepted German signals ordering the dismantling of the invasion fleet.

Bombing by night now became the main Luftwaffe strategy and the British had little active defence against it until the development of aircraft-based radar for their night fighters. The Battle of the Beams, in which the defeated Nazi Knickebein ELINT radio navigation guide for German night raiders is well documented in Dr R.V. Jones' book, *Most Secret War*. In it he describes how he detected a signal beamed from Germany that their bombers used to guide them to their targets in bombing Britain's cities by night. Jones was able to bend the beam so that German bombers dropped their bombs in open countryside. In a speech in the House of Commons, Winston Churchill paid homage to British

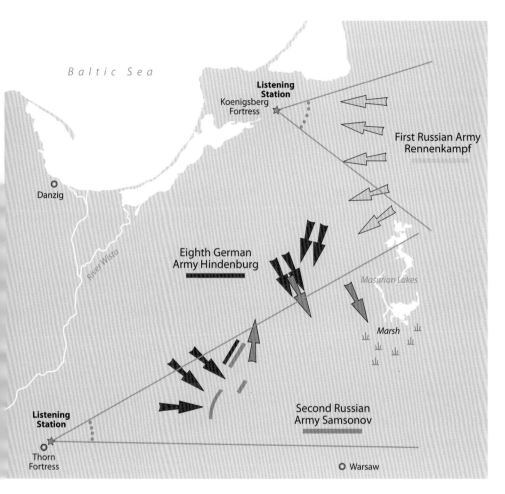

Baltic Sea

Listening
Station
Koenigsberg
Fortress

First Russian Army
Rennenkampf

Danzig

River Wisła

Eighth German
Army Hindenburg

Masurian Lakes

Marsh

Listening
Station

Second Russian
Army Samsonov

Thorn
Fortress

○ Warsaw

Hindenburg's crushing victory on the two Russian armies at the Battle of Tannenburg decided the shape of the First World War and even contributed to the revolution in Russia. The misuse of the Imperial Russian Army's new and largely untried wireless communications system enabled Germany to inflict one of the most complete defeats in military history.

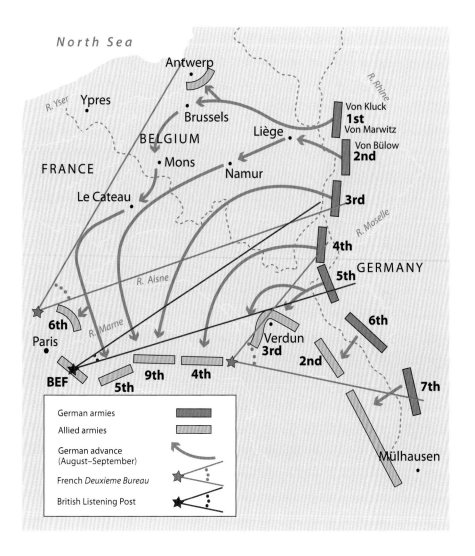

In French mythology the Battle of the Marne (1917) is known as the 'Miracle of the Marne'.
However, it was really the skill of the French in intercepting and decrypting the German coded messages that turned the tide of battle for the Allies.

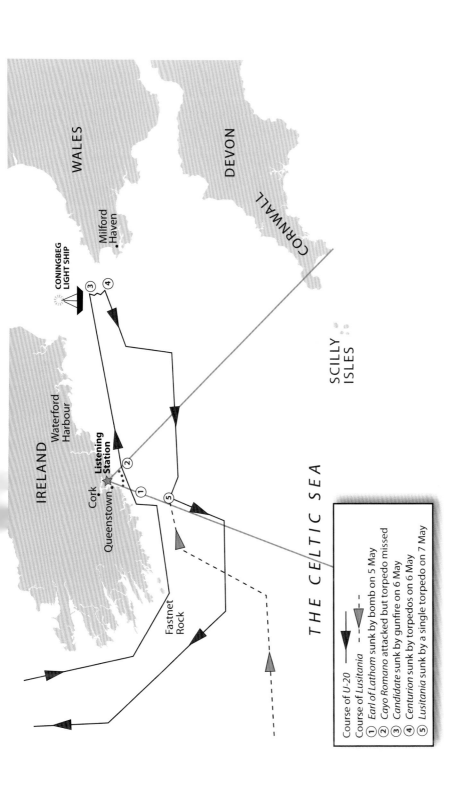

WALES

DEVON

CORNWALL

Milford Haven

CONINGBEG LIGHT SHIP

③

④

IRELAND

Waterford Harbour

Listening Station

Cork

Queenstown

②

①

⑤

Fastnet Rock

THE CELTIC SEA

SCILLY ISLES

Course of U-20 ——▶
Course of Lusitania – – –

① *Earl of Lathom* sunk by bomb on 5 May
② *Cayo Romano* attacked but torpedo missed
③ *Candidate* sunk by gunfire on 6 May
④ *Centurion* sunk by torpedos on 6 May
⑤ *Lusitania* sunk by a single torpedo on 7 May

The sinking of the *Lusitania* by a U-boat in 1915 could have been avoided if the signals transmissions had been more organised and timely. Conspiracy theorists suggest that the sinking was orchestrated to bring America into the First World War but this seems unlikely.

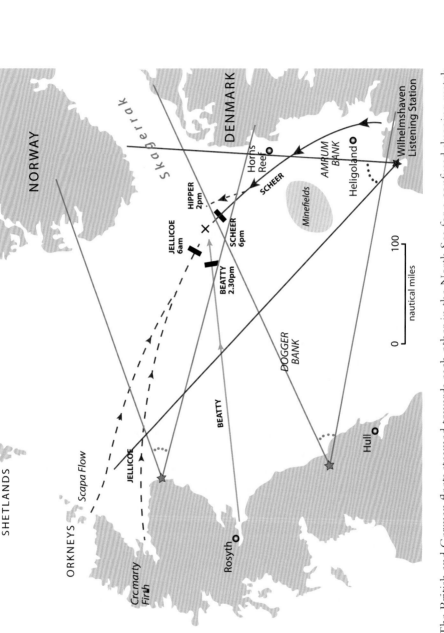

The British and German fleets steamed towards each other in the North Sea for a confused slugging match known as the Battle of Jutland in 1916. Both sides claimed victory, although very little was decided until the German fleet surrendered in 1918.

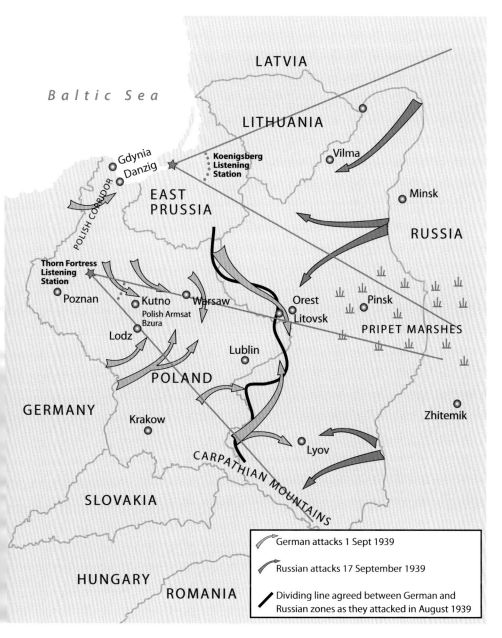

LATVIA

Baltic Sea

LITHUANIA

Gdynia
Danzig

Koenigsberg
Listening
Station

Vilma

Minsk

POLISH CORRIDOR

EAST
PRUSSIA

RUSSIA

Thorn Fortress
Listening
Station

Poznan

Kutno

Warsaw

Orest
Litovsk

Pinsk

PRIPET MARSHES

Polish Armsat
Bzura

Lodz

Lublin

POLAND

GERMANY

Krakow

Zhitemik

Lyov

CARPATHIAN MOUNTAINS

SLOVAKIA

HUNGARY

ROMANIA

German attacks 1 Sept 1939

Russian attacks 17 September 1939

Dividing line agreed between German and
Russian zones as they attacked in August 1939

The invasion of Poland was the trigger for the Second World War and a showpiece for Germany's new talent for mechanised warfare. The Germans and the Russians attacked Poland almost simultaneously and divided the country between them. Polish cryptographers escaped into Romania with their materials to continue the signals intelligence work at Bletchley Park.

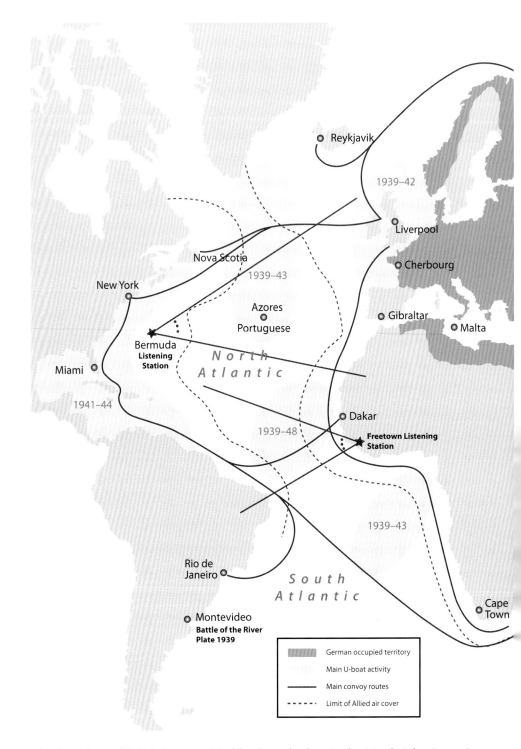

Reykjavik

1939–42

Liverpool

Nova Scotia

Cherbourg

1939–43

New York

Azores

Portuguese

Gibraltar

Malta

Bermuda
**Listening
Station**

*North
Atlantic*

Miami

1941–44

Dakar

1939–48

**Freetown Listening
Station**

Rio de
Janeiro

*South
Atlantic*

1939–43

Cape
Town

Montevideo
**Battle of the River
Plate 1939**

▨▨▨▨	German occupied territory
	Main U-boat activity
——	Main convoy routes
----	Limit of Allied air cover

The first phase of Britain's most critical battle took place in the North Atlantic, costing many
ships in its early stages. Bletchley Park was able to intercept transmissions from the German
U-boat fleet when an Enigma machine was captured by HMS *Bulldog* from *U-110* in 1941. This
enabled 'The Park' to direct convoys from the vicinity of known U-boats.

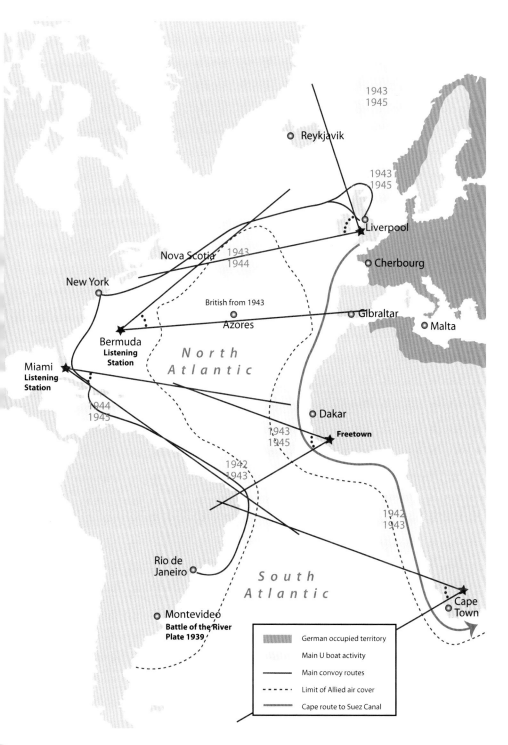

1943
1945

Reykjavik

1943
1945

Liverpool

Nova Scotia 1943
1944

Cherbourg

New York

British from 1943

Azores

Gibraltar

Malta

Bermuda
Listening
Station

North
Atlantic

Miami
Listening
Station

1944
1945

Dakar

1943
1945

Freetown

1942
1943

1942
1943

Rio de
Janeiro

South
Atlantic

Cape
Town

Montevideo
Battle of the River
Plate 1939

German occupied territory

Main U boat activity

Main convoy routes

Limit of Allied air cover

Cape route to Suez Canal

The second stage of the battle was won by the Anglo-American navies with the use of centimetric radar detection in conjunction with new weapons used by both warships and aircraft. As a result, U-boats were withdrawn from the Atlantic in May 1944, just in time for the D-Day fleet to operate without hindrance.

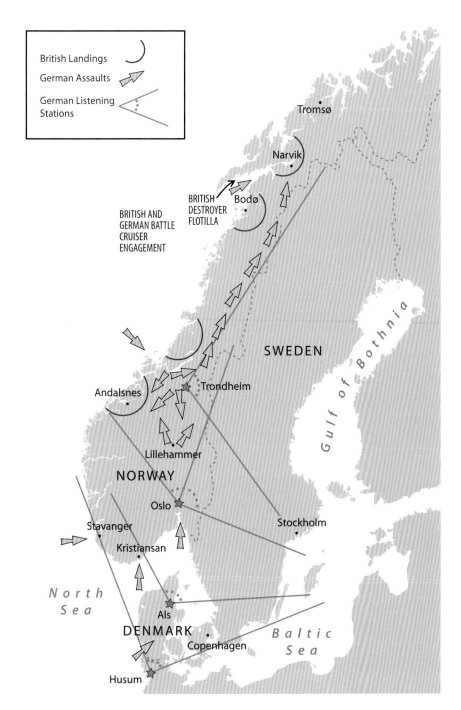

Tromsø

Narvik

BRITISH
DESTROYER
FLOTILLA

Bodø

BRITISH AND
GERMAN BATTLE
CRUISER
ENGAGEMENT

SWEDEN

Andalsnes

Trondheim

Lillehammer

NORWAY

Oslo

Stockholm

Stavanger

Kristiansan

Als

Gulf of Bothnia

*North
Sea*

DENMARK

Copenhagen

*Baltic
Sea*

Husum

The first engagement of British and German forces in Norway in April 1940. The British Army had to be evacuated after the operation failed, but the Royal Navy damaged the German surface fleet so severely that they were unable to intervene during the evacuation of Dunkirk. Signals intelligence played little part in this campaign due to bad reception in Norway's mountainous terrain, though the Germans did keep a close radio watch on Sweden for her reactions during the campaign.

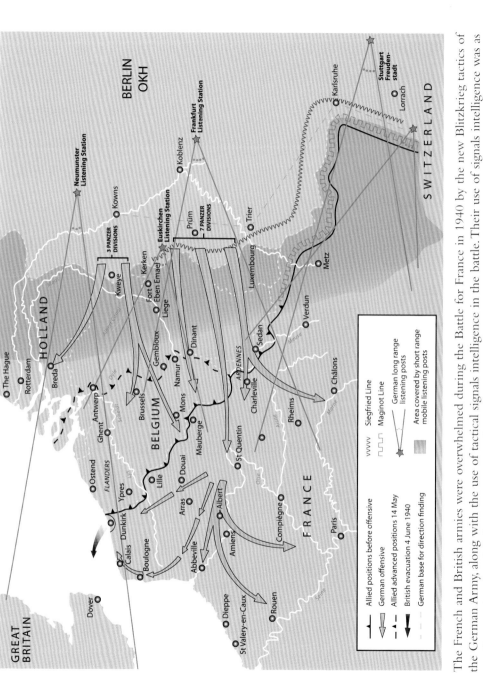

The French and British armies were overwhelmed during the Battle for France in 1940 by the new Blitzkrieg tactics of the German Army, along with the use of tactical signals intelligence in the battle. Their use of signals intelligence was as excellent as the rest of their manoeuvres in the campaign.

The German Army landings for Operation Sealion in 1940 were planned on a broad front in southern England, though the Kriegsmarine wanted a narrow one as they did not have the ships to support the army plan. Four German listening stations with experience of British transmissions were ordered by the OKW to advance to the French coast as the fighting subsided. They were to initiate radio intelligence against the British Isles and identify British Army units and their locations.

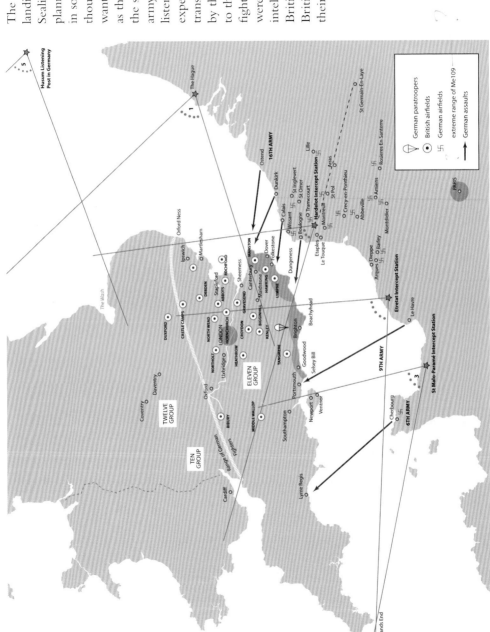

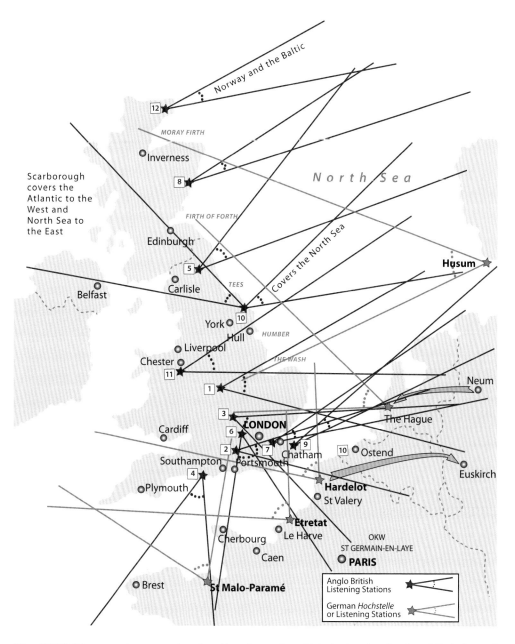

The SIGINT network set up by the Abwehr against Britain in the Battle of Britain and Operation Sealion was effective, although the Germans only used four listening posts to cover the whole of the UK. The British listening network was more complex, intense and specifically targeted.

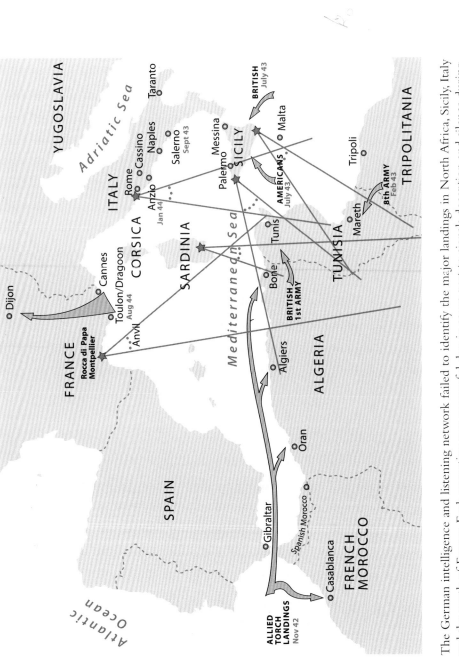

The German intelligence and listening network failed to identify the major landings in North Africa, Sicily, Italy and the south of France. Each operation was successful due in some part to signals deception and silence during the operation.

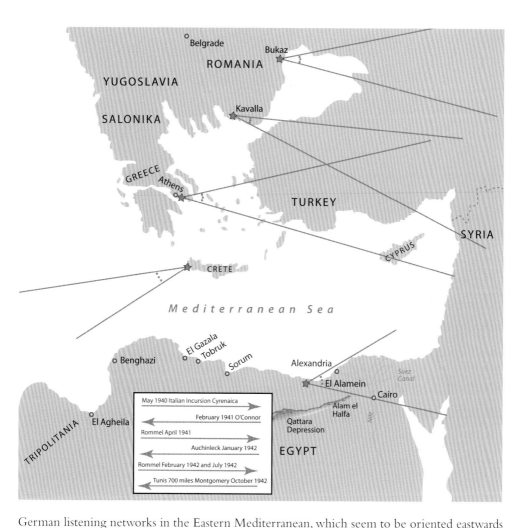

German listening networks in the Eastern Mediterranean, which seem to be oriented eastwards towards the southern Russian battlefields rather than Egypt and the Desert War. The American leakage of British Army plans and positions intelligence was covered by Rommel's mobile listening post, but this was captured just before El Alamein. Listening posts at Lauf repeated American transmissions for Hitler's attention. The advance and retreat of the Desert armies shows the nature of the campaign until signals intelligence was stopped.

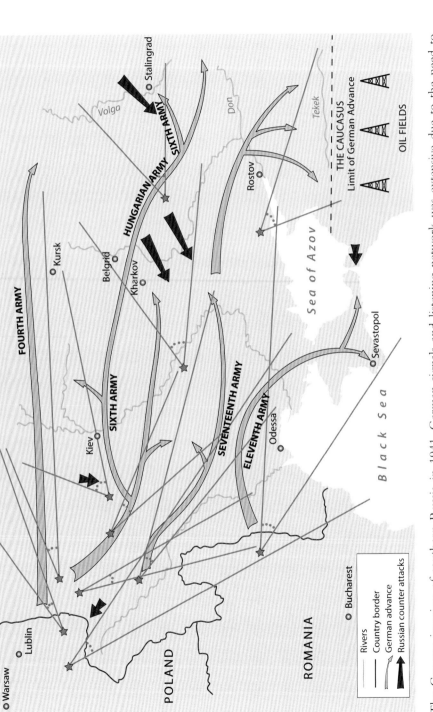

The German invasion of southern Russia in 1941. German signals and listening network was extensive due to the need to communicate over the huge distances involved. Their signals intelligence war was as complicated as the rest of this huge military campaign. Counter intelligence radio played a major part in directing the Russian partisans who later helped to destroy the German Army Group Centre in 1944.

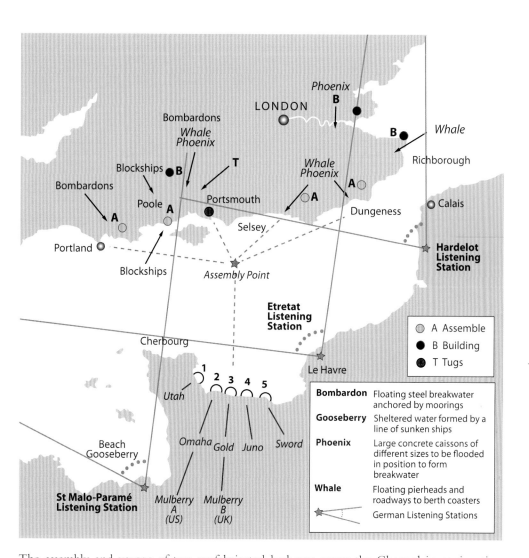

The assembly and voyage of two prefabricated harbours across the Channel in conjunction with the D-Day landings was one of the greatest engineering feats of the war. The way that the Allies managed signals traffic before and during this operation was masterful and derived from lessons learned in the Mediterranean. The withdrawal of German listening posts as the landings occured was well planned.

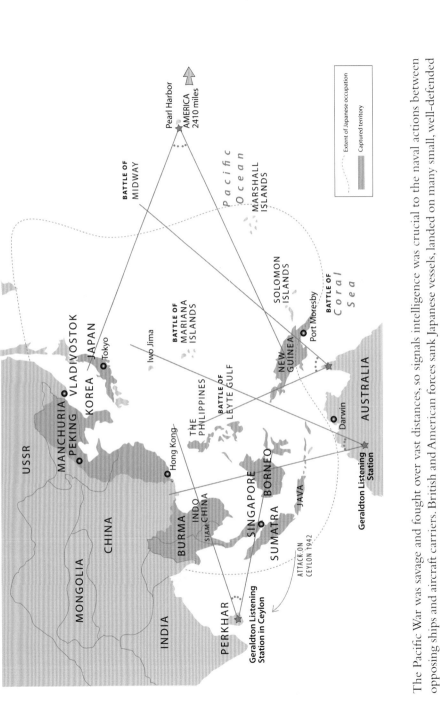

The Pacific War was savage and fought over vast distances, so signals intelligence was crucial to the naval actions between opposing ships and aircraft carriers. British and American forces sank Japanese vessels, landed on many small, well-defended islands, and battled through dense jungle before the atom bomb was dropped. The border between the Soviet Union and Manchuria, which Japan had occupied, was the subject of the Russo-Japanese Non-Aggression Pact in 1941. This led to a major development of the global war.

achievements in electronic warfare, and its companion ELINT. In his book on the Second World War, he wrote:

> During the human struggle between the British and the German Air Forces, between pilot and pilot, between A.A. batteries and aircraft, between ruthless bombing and the fortitude of the British people, another conflict was going on, step by step, month by month. This was a secret war, whose battles were lost or won unknown to the public; and only with difficulty comprehended, even now, by those outside the small scientific circles concerned [...] Unless British science had proved superior to German, and unless it's strange, sinister resources had been effectively brought to bear in the struggle for survival, we might well have been defeated, and, being defeated, destroyed.

Intelligence achievements were made in the air war with the main advances in ELINT techniques using photo reconnaissance and, in particular, radar. The German Knickebein beam transmitters were identified in an Ultra decryption after hints from captured aircrew were overheard on hidden microphones. Most orders and reports for aircraft operations during the air battles over Britain were carried on land lines rather than by radio transmissions so could not be intercepted. Photo reconnaissance by the RAF was able to identify the assembling of craft in French and Belgian harbours as preparation for the invasion of Britain. Other photos confirmed that those preparations were being dismantled and the attempt was not to be made. Aerial photography was not as easily available to the Germans, as the British were beginning to gain command of the air. Photos of the assembly of bogus army equipment and stores on the Essex and Kent coasts to mislead German intelligence into thinking there would be landings in the Pas-de-Calais area were not easily come by. Aerial photographs by the RAF were more plentiful and often directed at identifying and measuring damage to German industrial plants and capacity. However, photographing damage to factories and industrial plants as a way of assessing economic wounds inflicted by bombing on industrial production was an uncertain way of measuring it. A Mercedes-Benz plant at Sindelfingen, which was a principal manufacturer of aero engines during the war, was targeted by the RAF and bombed. Aerial photos showing that the roof had been blown off and the factory unusable were misleading. Factory managers used the cellars of the factory to rescue and reinstall undamaged machinery and to create an underground manufacturing plant. The building appeared to be a total ruin but by the end of the war it had doubled its output of aero engines in the sub-terrain premises. German industry showed great initiative in maintaining production in spite of destruction wrought by RAF Bomber Command. German industrial resourcefulness enabled her to double the production of tanks from its 1943 level to the end of the war but aerial photography evidenced many ruined factories and a different picture. Nevertheless Albert Speer, Hitler's Minister for Industry, whom the author stood guard over in Spandau Prison, after the war admitted that the bombing offensive slowed German production considerably but not as much as photo evidence showed. Identification of specific military establishments was quite another matter. The photographic identification of fortifications and gun emplacements before the D-Day landings were a considerable help to the planners for D-Day. The war's outstanding achievement of aerial photography was to identify the V2 rocket sites at Peenemünde and many other military targets were also identified by this technique.

Air Marshal 'Bomber' Harris used aerial photography to help justify his policy of mass bombing of German cities. His attempt to win the war by aerial bombardment of the German population rather than specific targets like oil installations cost the British war effort dearly and there is a fierce debate on the effectiveness of the bomber offensive. Was it a failure? And was it a defeat for the bomber crews or a triumph for them? The author saw the vast destruction in the major German cities just after the war. He also saw the official war cemeteries in Berlin with rows of white headstones marking the graves of a host of young British and American airmen who died in the offensive. If German radar networks had worked as well as the British 'Home Chain' version of fighter control system used in the Battle of Britain, aircraft losses would have been even greater. Radar was a revolution in air defence but Germany's radar version 'The Luftflotte' was designed for co-operation between aircraft and ground troops as used in the Battle of France with huge success. The radar control network did not prove so successful later when it had to detect and home German night fighters in on high-flying British bomber squadrons. Later German radar systems had to be improved to cope with the bomber offensive.

Radar was also about to be expanded into another application, as a weapon against U-boats and this was so effective that it led to the eventual demise of the submarine as a threat to Britain's maritime lifeline.

The Battle of the Atlantic

The war at sea was one that the Allies fought from day one until the end and could not afford to lose but it was a long and costly battle. The dimly remembered lessons learned in the previous war of the effectiveness of escorted Atlantic convoys in the battles between the U-boat fleet and the Royal Navy bore fruit. The first convoy left Canada three days after war had been declared consisting of thirty-six ships marshalled in rows of four vessels with escorts. Later convoys could consist of sixty ships covering many square miles of sea that sailed regularly from Canada's Atlantic coast to the ports of Britain. The Admiralty did make the mistake of letting vessels able to sail at over 15 knots to go unescorted in hostile waters, but that was corrected as the losses mounted. On any one day there could be 2,500 British merchant ships at sea, forming a juicy target for U-boats early in the war as there were not enough escorts. Often they would be poorly armed and with guns of too short a range to keep the U-boats at bay. This was known among submarine commanders as the 'happy time'. In the first four months of war to the end of 1939 German submarines had sunk 106 ships.

As war started the Kriegsmarine had only had twenty-six seaworthy submarines to operate in the Atlantic of which only about a third could be on station in their allotted patrol area, although another thirty-seven were undergoing trials before coming into service. The German *unterseeboote*, or U-boat, was a craft that a submariner would not describe as a true submarine as its hull was the shape of a ship and not the classical cigar shape of most submarine designs. The debate on the subject seemed to rather split hairs as the term (and threat) of the U-boat is well understood in the English language but virtually unknown in German. They use the word *unterseeboote* as a term describing any submarine of any nation. U-boats, as we will continue to describe the German submarine, were given low priority in production in the run up to and beginning of the war. Hitler listened to Admiral Raeder, who was an exponent of the powerful surface fleet rather than the submarine. The U-boats' spectacular successes in sinking ships, particularly the *Royal Oak*, in those early months changed all that. The U-boat fleet became Hitler's darling and the twenty-six roaming the Atlantic in 1939 had turned into 200 by the beginning of 1943.

The design of the sea-going U-boat, which was the one most used early in the war, was a simple serviceable one that could affect some of its own repairs at sea. They were sound vessels but the weapons with which they had to fight were not so effective. In the Norwegian Campaign, U-boat captains reported many torpedoes hitting ships and not detonating, and the track of the weapon then clearly indicated the position of the submarine. Work on the mechanism of the torpedo was urgently put in hand and the fault corrected in time to help the development of Germany's U-boat offensive in the Atlantic. To wage the submarine war they had built a number of types as the war progressed consisting of coastal submarines of 250 to 500 tons, mine-laying submarines of up to 500 tons and sea-going U-boats of up to 700 tons. The formidable ocean-going cruiser submarines of 1,000 tons or more did not come into production until very late in the war and too late to really affect the outcome.

The first sea-going U-boat was launched in 1936 and, subsequently, 1,170 were built; they mainly had the limitation of a range of operation of 6,500 miles at an average speed of 12 knots. Ocean-going boats were designed and built later and could cruise in mid-Atlantic at an underwater speed of 7 knots carrying twenty-two torpedoes but they took a long time to build. The fall of Norway and then France allowed the German U-boat fleet to establish bases on their Atlantic coast. In addition, the German *B-Dienst* was able to read the British convoy codes and foretell their convoy routings to waiting U-boats. Intelligence derived from these decoded signals was valuable and handled well by the British Admiralty Operational Intelligence Centre (OIC), based in the Citadel next to the Admiralty building and overlooking St James' Park. The British had a much greater investment in operators, plotters and senior officers directing operations to chart the deployment of German naval vessels. Also the OIC had the huge advantage of intelligence interceptions and evaluations from Bletchley Park. Even so, the Germans had their successes, particularly in the early stages of the war at sea with relatively less effort in plotting and controlling the battle. In the autumn of 1940 Admiral Dönitz began using the 'Wolf Pack' method of hunting for shipping – conceived in the First World War when the admiral was a young U-boat captain – and this was extremely effective. A line of up to twenty or so U-boats would wait across the line of an expected convoy. When it was sighted by one of the pack they would call in the others and attack the merchant ships together and overwhelm the escorts with a mass attack. Consequently sinkings rose dramatically in the Battle of the Atlantic, with thirty-six vessels being sunk in just two convoys. The devastating attacks continued as the German U-boat fleet grew to its peak of 330 operational submarines. The stranglehold tightened as over 200 ships were torpedoed, many of them off the east coast of America, until in June 1942 the highest number of sinkings in the war was recorded in Admiralty figures as 834,000 tons of Allied ships being destroyed.

By mid-1943 the tide of battle gradually turned in the Allies' favour due to a number of factors including better weapons and having broken Enigma again after months of not being able to read U-boat transmissions. The main factor, however, was due to the development and use of new aspects of radar. The results of British research became clear to the Kreigsmarine's *B-Dienst* listening station at Neumünster. They picked up a message about a large Allied convoy in the Atlantic and thirty U-boats were ordered lay in wait. Suddenly Neumünster picked up another transmission, instructing the convoy to change course to the north because a pack of submarines was in its path. It was first thought that the British had solved the U-boat cipher so a stand-by cipher was used to instruct the U-boats to take up positions on the new route to intercept the convoy. Neumünster then intercepted a further message the next day ordering the convoy to change course yet again, giving the exact position of the thirty U-boats. The Kriegsmarine was dismayed as it seemed that the German submarines were no longer invisible and protected when submerged. The submarines were ordered to move to new positions yet again. The convoy was redirected away from the U-boats for the third time and the locations of the submarines given again; and then a fourth time. Neumünster began to intercept an increasing stream of messages giving the locations of U-boats at sea, ten submarines in this area, twenty in that and three returning home, while seven were putting to sea. At the same time sinkings of Allied ships dropped alarmingly from 100,000 tons per operation trip of a U-boat to 10,000 tons and the losses of their submarines began to rise. The war at sea was going to be different from now on as the German submarine no longer had surprise on its side and the Germans did not know why.

The answer was a shorter wave of radar beams searching out enemy submarines more accurately but, more importantly, with a transceiver which was lighter and more portable. Long wave (but shorter range) radar together with Ultra had served the Royal Navy well in operations such as Cape Matapan in 1941, when Admiral Summerville's fleet surprised and severely damaged the Italian fleet. German radar development had also showed promise in the war's early stages but the balance was changing in considerable measure as Allied research created new weapons and procedures. The Kriegsmarine was conservative in its attitude to an emerging radar technology as they saw its main application as fixing locations of transmissions in a similar way to the use of D/F techniques. Its use was actively discouraged among young officers, who were taught that they had the superiority in German optical equipment (which was true) and they did not need knowledge of radar. However, centimetric and HD/DF radar, which was a British discovery, was about to make its debut. Radar had been in existence before the war but the sets were too bulky and the wavelengths used too long. The answer was to shorten the wavelength to 10cm or less to give a much clearer image of the target. This smaller set was fitted into British escort carriers and Coastal Command aircraft so the U-Boat fleet was unprepared for what was yet to come. Admiral Dönitz's steeply rising losses were also as a result of the development of new weapons, such as the hedgehog depth charge which did not explode until it hit a U-boat, and the new tactics of employing hunter groups of British warships. As a result, in the spring of 1943 fifty-six U-boats had been sunk, out of sixty-one that had been sent on patrol; such losses could not be sustained. Admiral Dönitz withdrew all his U-boats from the North Atlantic in May, virtually marking the end of the Battle of the Atlantic. There are no tombstones row on row for the 30,000 men of the merchant marine men that died; the memorial to them is to be found on Tower

Hill in London. On this memorial each ship's name is cast in bronze and by her name those seamen that went down with her. As Kipling wrote: 'If blood is the price of Admiralty, Lord God we have paid in full.'

Raiding Europe

The withdrawal of British troops from the European mainland after Dunkirk gave them the option to select targets for raids along its coastline. To do this they created the role and image of the Commando, not only to attack selected enemy installations but also to instil a fear of these almost mythical fighting men and where they would strike next. This certainly had the desired effect on German soldiers but produced a backlash from Hitler who decreed that any Commandos that were captured should be shot. Instilling fear in German troops of an enemy that would strike and fade away was quite successful, but any raid's objective, like any military operation, needed to be focused. Although to an outsider their purpose may not have been apparent, there was invariably a reason. Some raids were designed to deny a facility to the enemy with which to launch an attack on Britain or her shipping; others sometimes had a more esoteric reason and contained 'Snatch Squads' to seize anything that might be useful. This was a trick that the Allies learned from the German Abwehr, who first used it in Poland. The brilliantly conceived raid on St Nazaire on the coast of German-occupied France did not contain such squads but its obvious objective was to destroy a major docking facility suitable for berthing any one of Germany's few major warships. Another more low-key operation was an attack on the Lofoten Islands off the coast of Northern Norway. The Royal Navy was briefed to 'pinch' any Enigma material that they could lay their hands on in a raid with a cover purpose of destroying infrastructure on the island. They managed to find a set of Enigma rotor keys and documentation on a German tug but the machine had been thrown overboard. Later a couple of German trawlers in Icelandic waters whose purpose was to transmit weather reports back to base were pursued by a British cruiser sent to intercept them. They ditched their precious machine in deep water but some helpful documents were found.

One cheeky raid was made on Bruneval, on the French coast, in February 1942. With the code name Operation Biting, it had the objective of capturing German Würzburg radar equipment from an emplacement on the coast near Le Havre. German radar defences were causing British Bomber Command increasing losses of aircraft and scientists needed to know how the enemy's system worked so they might devise counter-measures. German Luftwaffe radar systems were used to detect British bombers with the Freya-Meldung-Freya network that had two complimentary halves. The first was Freya, which was a long-range radar system but not very exact in its plot of the target aircraft, and Wurzburg which was a shorter range system but much more accurate. In co-operation, they could pinpoint an enemy aircraft and guide German night fighters onto an individual bomber. Snatching the Würzburg radar array was the target for the Commando team, with scientific advisors arriving at Bruneval by parachute and landing craft. After capturing the site, technicians dismantled key pieces of the installation and brought them back to England on a landing barge for scientists to study. The technical knowledge gained of Germany's radar system along with captured radar technicians enabled British scientists to devise the 'window' radar-jamming system, which was just strips of aluminium foil of a specific dimension scattered by the bombers to blind German radar systems. One consequence of the Bruneval raid was that the Germans improved defences of all Würzburg sites, making them more visible from the air and, therefore, easier to bomb.

A raid on the German-held port of Dieppe in August 1942 has, until recently, been shrouded in mystery. Carried out by 6,000, mainly Canadian, troops, including a tank regiment with the support of 300 ships and 800 aircraft, it was the biggest raid at that time. Six months before the Dieppe Raid, the Kriegsmarine had added another rotor to the Enigma machine used in their naval codes, creating an almost unbreakable code used by U-boats at sea. Consequently, Bletchley Park had been unable to decrypt coded signals from Germany's submarine fleet in the Atlantic, at great cost to British shipping. The raid was a costly failure because, when approaching the port, the British fleet had 'bumped' an enemy coastal convoy and the exchange of fire lost them the element of surprise. The Dieppe garrison was alerted and gave a good account of themselves, repulsing the Canadians with bloody losses. Both German and British radio intelligence learned some lessons from the operation. The preparations for the raid were secure mainly because a strict radio silence was maintained by the attackers. Radio signals sent during the fighting were received as far away as the Hague Fixed Intercept Station and fainter ones by Étretat in France but neither were able to determine the source of the transmissions and thought that fighting had broken out in Holland. The main reason for German confusion was the long chain of command passing through St Germain in Paris before being passed to OKW West with all the consequent delays. British transmissions increased during the afternoon of the raid so that the situation became clearer, but by the time arrangements had been made for a withdrawal it became obvious to the Germans that the assault was only a raid. Changes were subsequently made to the German Army reporting system, though the Canadian code words, such as colours for specific beaches, were difficult to penetrate.

The mystery objectives of the raid was revealed only recently by David O'Keefe, Professor of Military History at John Abbott College in Montreal. After studying contemporary papers and documents about the raid, it became obvious that a snatch squad were after an Enigma code machine. One of the leaders of the raid was Captain Ryder RN, who had distinguished himself at the St Nazaire Raid. He had orders to penetrate the harbour defences and storm the headquarters of the German garrison with 40 Commando of the Royal Marines who had been trained as a snatch squad. The Hotel Moderne in the centre of the town was the German garrison nerve centre and would most certainly have had an Enigma machine, which was the prize the marines sought. The Marine Commandos attempted to enter the harbour several times but were repulsed by German fire from gun emplacements, as were most of the Canadian forces on the beaches. The Enigma snatch operation had failed and so did many other elements of the operation, but this taught the D-Day planners a number of lessons concerning amphibious landings on a hostile and well-defended coast. Dieppe was a costly lesson; chief among them was the need to think good and hard about how to establish the use of an operational port on the enemy coast. It was from this early stage that the concept of the Mulberry Harbour (a portable harbour) was born. The raid also had another important strategic lesson for the Allies: an assault on the coast of France were going to be much more difficult than at first thought, at least by the American military. The US Army and, principally, General George Marshall had been enthusiastic about attempting a landing on the French coast at Pas-de-Calais as early as 1942, but Churchill saw the difficulties of a seaborne invasion. The cost of Dieppe made the Americans reconsider and plan the landings more meticulously than they had done previously.

The Allies' lack of success in the raid convinced the D-Day planners that there was no easy way to capture a French port so they started to plan the artificial harbour codenamed Mulberry. The author worked in the offices of Pauling & Co., who were civil engineers and one of the organisations concerned with designing, building and towing the artificial harbour across the Channel and installing it on the shoreline of the quiet French seaside resort of Arromanches. The components of the two harbours were codenamed Bombardons that were floating steel breakwaters anchored to moorings to provide calm water. Whale was the code for the floating pier heads and roads that would bear a heavy tank and lastly Phoenix were the huge hollow concrete caissons that could be positioned and then flooded to sink to the sea bed and form a great breakwater. The harbour was a most impressive sight with the constant coming and going of ships unloading stores and armaments to go to the front. Veteran soldiers who worked on Mulberry often came to Pauling's offices and recounted the only relaxation they had in those hectic days after its installation. They would sit and watch the faces of German prisoners as they were brought through the sand dunes to behold a large functioning and frantically busy harbour where there had been none days before.

Meanwhile, the British still desperately wanted an Enigma machine but did not have long to wait. The U-boat 110 attacked a convoy south of Greenland in early May 1941 and the British destroyer HMS *Bulldog* damaged it, forcing it to the surface. Kapitänleutnant Lemp commanding the submarine set explosive charges to sink his vessel as his crew abandoned ship but the charges never went off. A party of British sailors led by Sub-Lieutenant Balme boarded her and seized a whole treasury of secret papers and an undamaged Enigma machine. Balme was awarded a well-deserved Distinguished Service Cross and the machine and its attendant library of documents was sent back to Bletchley Park for analysis and usage. It was the beginning of the Park's authority as a code-breaking and intelligence centre, which it retained to the end of the war. Capturing the Enigma machine and breaking its code resulted in the chain of surface ships supplying not only U-boats but also warships being cut. The Milch Cow submarines (the nickname for the Type XIV U-boat, whose only weaponry was anti-aircraft guns) carrying supplies to sea-going U-boats was now the only option for Dönitz. It was also a severe limitation on German warships wishing to break out into the Atlantic to raid British shipping. The tankers and supply ships which they would use were all identified, located and sunk within a month of HMS *Bulldog* capturing the Enigma machine.

Germany's Surface Fleet

The grand strategy of the war for the German leaders was based on dominating mainland Europe with the Luftwaffe supporting Hitler's hard-hitting Panzer columns. The Kriegsmarine was expected to play an important supporting role in the operations against a superior Royal Navy. A limited share of money and resources therefore went to the Kriegsmarine of which the submarine service got an increasing share as a reward for its successes in the sea war. The result of this policy was that the German surface fleet consisted of six capital ships, one of which was ancient in naval terms, six cruisers and twelve destroyers – hardly an armada to face the British Home Fleet in its own waters. The first of Hitler's heavy warships, the *Graf Spee*, was reduced to a smoking ruin in the first weeks of the war, followed by the action in Norwegian waters costing the Kriegsmarine the heavy cruiser *Blücher*, two light cruisers and ten destroyers, in addition to which three capital ships were damaged, one of them so seriously that it took a year to repair it. The losses that the Kriegsmarine suffered were severe but the capital ships that were left still constituted a threat to the jugular vein of the British in the Atlantic if they came out of harbour, which they did from time to time.

This happened with the pocket battleship *Admiral Scheer* in 1939 which sighted Convoy HX-84 while patrolling the Denmark Strait off the coast of Iceland. Escorts to shepherd British convoys from Halifax in Canada's Nova Scotia to Liverpool or Belfast were scarce at that time and a convoy could sometimes have a single armed merchant cruiser as their only escort. Armed merchant cruisers were just large merchant ships with a few guns installed and manned by naval personnel; but even guns were in short supply early in the war. The author's father's ship HMS *Montclair* was just such a vessel and some of the guns that were meant to scare the U-boats were wooden mock-ups. The escort to the Icelandic convoy was one of these vessels and commanded by a friend of our family, Captain Edward Fegen RN, who had the nickname Fogerty Fegan due to a voice that could double for a fog horn. The 11in guns of the *Admiral Scheer* opened up so Fogerty immediately ordered the convoy to scatter and drove his ship at full speed towards the German battleship while making smoke to cover the escape of the ships in the convoy. It took time for the *Admiral Scheer* to destroy Fogerty's ship, allowing the convoy to disappear into the northern mists and the smoke. Only five ships of the convoy were sunk as a result and Captain Fegan was awarded the Victoria Cross.

During the war only two of Germany's capital ships faced the Royal Navy with guns blazing, most of the rest were destroyed or badly damaged while still in dock, suffering from the attentions of Bomber Command. The first to go was the battleship *Graf Spee*, whose dismal end has already been described, and the second was the *Bismarck*. One of the biggest warships in the western hemisphere, she could out-run or out-gun almost any one of the Royal Navy's capital ships, most of which were ageing vessels and several of which had fought at Jutland almost twenty-five years earlier. The *Bismarck* was, therefore, the principal target for the Royal Navy and, when she and several attendant cruisers broke out into the North Sea at the end of May 1941, all the heavyweights of the British fleet began to search and give chase. The German squadron was sighted on 24 May and the Royal Navy's battleships began to close in on the German squadron. After a few exchanges, the pride of the Royal Navy, the battleship HMS *Hood*, was hit in her ammunition magazine and disappeared beneath the waves in three minutes. Only three survived out of her crew of 1,500 but the *Hood* was not the only one that was hit. A shell had damaged the *Bismarck*'s fuel supply and she was now short by 1,000 tons of fuel. In addition an air attack scored a torpedo hit but she had given the gathering fleet of British warships the slip in the night. Where was she going? Would it be a German or Norwegian port or would it be a French one?

The answer was about to become clear.

The German Press reported the sinking of the *Hood* and the battle of the giant warships out in the Atlantic. The battle was followed with great enthusiasm by the German media and public but even they would have realised the numerical odds were against their flotilla. The Royal Navy's finest began to gather around the *Bismarck* so she had to run for port. One spectator was a high-ranking Luftwaffe officer in Athens whose nephew was a midshipman on the pocket battleship. An urgent wireless message using a diplomatic code which turned out to be even less secure than that of the GAF (German Air Force) cipher was sent from Athens asking after the boy's welfare and safety, and incidentally for which port *Bismarck* was headed. Presumably a Luftwaffe plane might fly down to meet his relation. The reply came back that the *Bismarck* had set course for the port of Brest with all possible speed. Bletchley Park picked up the message and immediately relayed the information about the German intentions to Admiral Tovey, who was directing search operations for the fugitive. Some controversy exists as to the effect of the above decrypt, but either way she was about to be spotted.

A Catalina flying boat on the 26th at 10.30 a.m. confirmed the warship was 700 miles west of Brest and steaming for home at full speed. The British fleet was closing, but not fast enough to catch her before she came under the protective wing of the Luftwaffe. The British fleet had to slow her down, so that evening the aircraft carrier HMS *Ark Royal* launched an air strike on the *Bismarck* in atrocious weather. A squadron of Swordfish biplanes loaded a torpedo each and took off to make a courageous attack on the heavily armed warship. They scored a couple of hits, the most important of which was on the rudder of the battleship, which jammed it. The great vessel started turning in uncontrollable circles and was trapped. During the night the British main hunting force arrived and methodically reduced the *Bismarck* to a battered and burning hulk. A final blow by the cruiser HMS *Dorsetshire's* torpedoes saw the *Bismarck* sink with her battle ensign still flying along with most of her crew, including the midshipman whose uncle may have caused it all.

Other German warships did not come out to face the Royal Navy, but that did not mean that they did not pose a great threat to the merchant fleet plying the Atlantic. At the centre of that threat was the pocket battleship *Tirpitz*, a warship almost as big as the *Bismarck* and based in Le Havre which received constant visits from the RAF. Hitler decided it should, with its attendant flotilla of cruisers and destroyers, move to another berth and that move should be an audacious one. The *Tirpitz*, accompanied by heavy cruisers *Scharnhorst* and *Gneisenau*, would make a dash through the English Channel, past the white cliffs of Dover in daylight to a safer berth in the fjords of Norway. The success of the Channel Dash, as the British press called it, was due to the failure of the British aircraft that regularly patrolled the German warships in harbour. Their radar sets had failed and the absence of the German warships did not register with them. Land-based radar stations scanning the Channel approaches were jammed by the Germans at the same time. The fleet passed through the Straits, although not unscathed, as both the *Scharnhorst* and *Gneisenau* were damaged by mines, and had found a haven but not a safe one from the RAF in German ports. The *Tirpitz*, hidden in the Norwegian fjords, remained a threat to British shipping into the autumn of 1944 when it was heavily attacked a number of times in Altenfjord, Norway, by carrier-based aircraft during July and August. The reason for the attack was that from August 1941 over forty convoys containing over 800 ships had been sailing to Russia with equipment and arms to support their fight against the German invader. Every trip was a bitterly fought battle, in particular for convoy PQ 17 which lost thirty-four of its thirty-six merchant ships sunk. The threat of the *Tirpitz* coming out to attack caused the convoy to disperse, losing the protection of their escorts, so while the German warship was there she constituted a threat to British shipping. She had to destroyed by British bombers, and she was.

The balance of power at sea was absolutely essential to the British, whose battle fleet was spread across the Atlantic, the Mediterranean and increasingly the Pacific as the Japanese threat began to emerge. The Italians had a surface fleet consisting of five battleships, fourteen cruisers and twenty-seven destroyers all based in the port of Taranto in Southern Italy. Following the invasion of Greece in October, the Axis Powers dominated the Aegean and Adriatic seas and threatened the island of Malta. Despite this, two British convoys fought their way through to resupply Malta under threat from the Italian fleet, although they never left port to intervene. An audacious raid by just twenty Swordfish bi-planes carrying torpedoes and bombs caught the Italians by surprise in Taranto harbour and sank half of their fleet. The remaining ships still left posed a threat, however, so when an Ultra intelligence evaluation from Bletchley Park warned Admiral Cunningham in Alexandria that the Italian fleet was about to put to sea to attack ships evacuating British troops from Greece, action had to be taken. The British fleet surprised them at Cape Matapan in the first fleet action to owe its success to signals intelligence, in which Cunningham's ships sank three cruisers and two destroyers. During the evacuation of British troops from Greece and Crete the Royal Navy lost several ships, but if the Italian fleet had been intact the losses would have been infinitely worse. The Royal Navy had been able to dominate the Mediterranean with the use of signals intelligence.

Germany's foes grew in strength as the war progressed but so did the unreliability of her allies, the chief among them being Italy. The bond between the two countries was created in 1940 based mainly on the personal association between Hitler and Mussolini. The 'Pact of Steel' they signed was a declaration of co-operation between them. A secret protocol in the pact encouraged a union of military policies including an exchange of intelligence intercepts of all kinds. There were two intelligence services in Italy similar to Germany's Abwehr and the SD. The *Servizio Informazioni Militare* or SIM was military intelligence while the *Opera Volantario per la Regressione Dell' Autifasismo* or OVRA was the Fascist Party's intelligence service. OVRA was a pale imitation of the Gestapo but they indulged in more international operations to harass and control anti-fascists in Yugoslavia and the Adriatic. OVRA even attempted to get anti-fascists sent back to Italy from Canada but this was rejected by the Canadian government. Both SIM and OVRA had their own signals intercept bureaus resulting in much the same unhappy rivalry and duplicity as experienced in Germany.

Mussolini's army was building Italy's North African Empire in Libya as Germany's war began so he declared war on Britain and France in June 1940. Marshal Graziani in Libya commanded almost quarter of a million men in the Italian Tenth Army. They advanced on the British in Egypt in September but got no further than Sidi Barrani, just over the border and almost 300 miles from Cairo. They set up camp and stayed there, scared by a signal intelligence deception that made the British force of 30,000 sound a lot bigger than it was. The Italians were then outwitted by the smaller British force that routed the Italian Army during a brilliant campaign, taking 130,000 prisoners and driving the rest 500 miles west into Cyrenaica. The British General O'Connor could probably have cleared the enemy out of North Africa, having greatly benefited from breaking the Italian Air Force ciphers. Unfortunately, Churchill ordered him to send a large part of his force to support the Greek Army as the Germans invaded. In the meantime, General Rommel and his *Deutsches Afrika Korps* arrived to bolster the Italians.

The Abwehr regarded the Italians as a security risk for several reasons, although their intelligence service performed quite well. Sadly for the two Axis leaders, the Italian Army, its General Staff and even the people of both countries did not share their leaders' respect for their 'allies'. Nor did the Germans trust the Italians, for even at an early stage in the war Admiral Canaris, as head of the Abwehr told, his staff that he believed that if matters in the war became critical Italy would defect to Germany's opponents. He added that it was not the place of the Abwehr to take any action in this belief but they must leave it to their political leaders to decide policy in international affairs. Canaris was right of course, the Italians began to weaken in their resolve to wage war as early as the autumn of 1941. Intercepted Italian signals traffic began to mention a separate peace at a time when the German Army was still advancing victoriously into Russia. Even at a time of German success, there was much opposition to sending Italian troops to Russia. Italy had been the reason for Hitler's changes in policy in Africa and the Balkans, none of which would prove beneficial to Germany.

The Balkans

The Italians, driven by Benito Mussolini, aspired to create an empire even though they lacked the economic and natural resources to build or maintain one. The Italians were not a naturally war-like nation, but Mussolini did not know it until they were faced with a serious war. Italy had misspent what wealth she had in the 1930s on ill-advised schemes and imperial wars, and now her armed forces were run down and equipped with obsolete aircraft and weapons. In spite of this Mussolini declared war on France on 10 June 1940 and invaded it from the south as it was collapsing under the onslaught of the German Army. Mussolini included Britain in that declaration as he viewed the British as militarily weak and also ready to collapse, but the German High Command saw things differently. The victories of the German Army had secured its western seaboard flank from the tip of Norway down to the borders of France and Spain. What the Italian entry into the war did was to open up the southern flank along the southern Mediterranean to the British forces based in Egypt and the Middle East. The German High Command believed that the British Army in the region was far stronger than they actually were, due to some masterly deceptive initiatives by General Wavell, C-in-C Middle East, and they feared an attack by him. From the perspective of the German General Staff in Berlin, the Italian entry into the war was a disaster. As if to confirm their view, the Italian attack on the British in Egypt was followed by an invasion of Greece with 200,000 men in late October – all with disastrous results.

Although it had some elite units, the Italian Army proved to be made up of largely incompetent soldiers and one of their principal failures was the way the Italians used their radios. Their wireless set-up betrayed the identity of virtually all units in the Italian Eleventh Army as each one had the same call-sign and did not change it for days at a time. Moreover, the Greeks had obtained the cipher systems of the Italian Army and in the first days of the Balkan campaign had been able to decipher much of the radio traffic of their opponents. This enabled them to outmanoeuvre the Italians and gain superiority in military actions, embarrassing Field Marshal Badoglio, Chief of the Italian General Staff, who sacked a number of his staff officers as a consequence. The Greeks assumed that Hitler would not let his allies be beaten, and they were right, so in addition to his invading Yugoslavia he added Greece to the list as well. The Greek government was under pressure and asked the British government for military aid so after much soul searching Churchill ordered Wavell to send 60,000 men. As a result, the German Army invaded Greece the next day. The British Army planned to build their defences in strong positions in the interior; this did not appeal to the

Greek government who wanted to defend the sacred soil of Greece at its frontiers. Hastily erected defences proved too weak for the Greek Army to hold, so they and the British forces were easily overcome by the German Panzers. British SIS agents began to infiltrate the Greek mainland even as the Germans occupied the country, but not always successfully. One of these was a Greek officer who was landed by submarine close to Athens to determine German Army strengths and was instructed to radio his controller in Cairo on landing to report his safe arrival, but the Abwehr were listening. They sent a faked reply asking him to switch to another frequency and promising that a submarine would come and pick him up; it was the Gestapo who welcomed him instead. A good many Allied agents were lost in the Balkans as the situation was very different to that in Western Europe, with the partisans often fighting each other almost as much as they fought the Germans. The role of the agents was largely diversionary and to keep Hitler's fear of an Allied landing in Greece alive. They kept a great many German soldiers tied up on the Greek coast waiting for the Allied landings to come.

The Greek campaign was not a total disaster for British troops but it was still a considerable reverse. The Germans had invaded from the north and it took three weeks to enter Athens but by then the British soldiers were withdrawing from the country to the island of Crete. General Freyberg, commander of the Allied forces on the island, had a mixed force of British, Australian and New Zealand soldiers, some of them evacuees from Greece. The Battle of Crete started with General Freyberg being told of the exact details of a German airbourne assault from Ultra signals intelligence. Freyberg's force was too small to take full advantage of his intelligence as he faced a ferocious airborne assault for ten days before evacuation was ordered yet again. Hitler had planned to invade Russia in May but the opinion of the German generals was that the friction, as they called it, in Greece had cost them several valuable weeks. These would be worth as many months of campaigning time to Hitler before the year ran out and the Russian winter was waiting.

Other Balkan countries were more quietly annexed as the German Army moved in, first as advisers and later more strongly as 'friends' had requested them to do so. King Carol of Romania showed himself to be anti-Nazi and was overthrown by the Iron Guard whose organisation bore a remarkable resemblance to the National Socialist SA in Germany. Bulgaria was promised that the cession of Greek Macedonia would be theirs if they co-operated in intelligence and other matters so a German intercept station was set up in Sofia. The station was given the code name of Borer and had to monitor diplomatic traffic in the area as well as Russian and British military transmissions. Now the German-Italian Axis Empire stretched from the Atlantic coast to the mouth of the Danube and there was only one fly in the ointment, or rather the hornet's nest of Yugoslavia, and the conflict was about to spread out to the Middle East.

The Middle East

With most of Europe in the bag, Hitler was planning his attack on the Soviet Union with the objectives of extinguishing Communism and obtaining oil in the Middle East; a commodity of which Germany was sorely in need. Wilhelm Flicke's papers show that up to this time radio traffic in the region had been of passing interest but now his intercept station was ordered by the OKW to prioritise the interception of radio traffic in Turkey, Iran, Iraq, Syria, Palestine, Trans-Jordan and Egypt. The Abwehr began to establish a propaganda and espionage network in Turkey and several of the Arab countries, as plans were made to attack southern Russia from the Balkans, either through Turkey or Crete and into Syria. A second option was through North Africa via Egypt, as the violation of Arab states' sovereignty were no obstacle to Hitler's plans. In Iraq a powerful pro-German group led by Rashid Ali al-Gaylani were preparing for a coup against the government and a takeover of the Mosul oil fields. In Iran Reza Khan, the father of the future Shah of Persia, was ready to act when German troops crossed either the Suez Canal or maybe even the Bosporus. There were even political allies in Afghanistan waiting for the signal to rise against the British. Hitler saw the need to use a religious approach to gain Arab co-operation so he had repeated meetings with the Mufti of Jerusalem in Berlin. A jihad against the British was promised throughout the Middle East when Germany was ready. No opposition was expected from Syria as it was a French mandate and administered by the French General Dents who was loyal to Vichy (the puppet German government in France). The only country seen to be friendly to Britain was Trans-Jordan because the ruler Emir Abdullah had contributed troops to fight with the British Army. However, his country was small and not seen as a problem and little resistance was expected there. The German Army was poised to invade Yugoslavia and then carry out an assault on Crete (as already described), but then things began to go wrong.

At the beginning of April 1941 the planned assault on Yugoslavia was launched at the same time as Rommel began his advance on Benghazi. A coup occurred in Iraq where the Prime Minister Nuri al-Said, who had fought with Lawrence of Arabia in the First World War, was deposed and killed. Rashid Ali al-Gaylani formed a new government and issued a proclamation expressing friendship for Great Britain and Turkey which no one believed. Strong British Army units immediately landed in Basra harbour and the Mosul oil fields which supplied the fuel driving the warships of the British Mediterranean Fleet. Without a constant supply of oil, Britain's warships would have been immobilised and the Italian fleet, although damaged at Taranto and Matapan, would have dominated the Mediterranean, allowing them to provide cover for German Army landings anywhere in the region. If the RAF and US Air Force had concentrated on German oil production facilities more exclusively then the war might have been over somewhat quicker.

Anti-British demonstrations and riots erupted in several Arab countries and German aircraft began to fly technical and military personnel into Syria. Rommel had advanced beyond Egypt's frontiers towards Cairo where Egyptian nationalists led by Colonel Nasser expected to welcome him. British influence and even its presence in the Middle East and oil fields was on a knife edge.

Meanwhile, German air attacks had begun on Crete in the middle of May but the assault was more costly than was expected with 16,000 dead and 400 aircraft lost due to a comprehensive interception by Ultra of their plans. The operation took more precious time than the German High Command expected and a quick victory did not follow what they thought was a surprise attack. Men and aircraft lost in Crete were intended for use in the occupation of Syria and while the battle for Crete was raging British troops crossed the border into Syria and Wavell's men calmed the rioting mobs in Cairo. Rommel was still at El Alamein but the rest of the picture in the Middle East had changed dramatically. The German High Command, which was in effect Hitler served by a few yes-men of the General Staff, were sure that their coding machines were secure and their codes unbreakable. The intelligence community was less convinced, as a direct quote from the papers of one of Germany's senior cryptographers Wilhelm Flicke clearly shows:

I have to conclude without a few words regarding the struggle for Crete which was a decisive factor in the collapse of the plan for a leap into the Near East. We (the Germans) had counted on taking Crete quickly in a surprise attack. The troops, aircraft and ships that were made ready were intended for the landing in Syria. Unexpectedly a large part were used up in the struggle for the Island of Crete.

How can we explain the unexpected resistance on this island? I can offer no proof of the assumption that I am to put forth, but all my observations indicate that the garrison had precise knowledge of the impending attack. The resistance of the Allied troops, which almost everywhere gave the impression that they had been waiting impatiently for the German attack, and the events off Cape Spatha where the German convoy was scattered and almost destroyed, admit of no other explanation than that the defenders of the island were fully informed by radio of all details. Possibly there were English radio agents in Athens who got their information from High German sources, but it is quite as likely that the German cryptographic system had been solved and that German traffic was being read. This question only the British can answer.

The Wilhelm Flicke papers

The TICOM investigation found that some senior German cryptographers believed that the Enigma code and maybe other codes could have been broken. Flicke never learned of the achievements of Bletchley Park as he died some years before the secrets of the Park were revealed. The extract quoted above was written in 1945 and several warnings of a similar nature were given to Hitler before then, but he preferred to ignore them. The Führer committed the ultimate sin of intelligence evaluation: only believing those things that agreed with his firmly held convictions.

The Desert Campaign

The campaign in the Western Desert along the shores of North Africa was not planned by either side. A relatively small formation of British, Australian and Indian soldiers were engaged in guarding the Suez Canal while further along the coast a much larger force of Italian troops were garrisoning their colonies in Libya, Cyrenaica and Tripolitania. The sudden confrontation as Italy declared war on Britain would create a battlefield of 600 miles of desolate sand and desert, interspersed with some small towns and villages. These were linked mainly by a coast road running from Alexandria to Tripoli through a largely empty, windswept, fly-infested space in between. Marshal Graziani led the Italian Army's first attack, almost quarter of a million strong, across a largely unmarked border into Egypt. They had advanced about 60 miles to Sidi Barrani and then stopped to fortify their position, having suffered light casualties at the hands of the British, but more importantly to defend their extended supply lines. The British intelligence community had begun to build up a tradition of effective military intelligence and deception of its enemy as the quality and volume of Ultra intelligence evaluation slowly improved. It would reach a level of excellence that would slowly permeate the British Army and serve its country well for the rest of the war.

A part of that improvement began with General Archibald Wavell who had a good schooling in military deception from General Allenby under whom he served in the First World War against the Turkish Army. During that war, Allenby masterminded several disinformation operations. One of them concerned a briefcase containing false plans lost in no-man's land (shades of the book and subsequent film, *The Man Who Never Was*). This misled the Turkish generals into thinking the British forces would attack in a particular place, but they then outmanoeuvred the Turks, enabling troops to capture some almost impregnable positions in Gaza with little loss of life. As a result of his experience under Allenby, Wavell did not have the fixed mind set of some senior officers of his generation concerning the use of intelligence. He was able to deceive the Italians into thinking Egypt's garrison was actually twice its size; this was done by maintaining a level of radio traffic that indicated a much larger British force. The ruse was the beginning of a series of disruptive and deceptive operations (or non-operations) that became a British speciality throughout the rest of the war. Techniques were increasingly developed in the desert to identify and exploit the enemy's fears through the use of deception. The most important deception was when General O'Connor convinced the Italians that he had a much bigger force than he had. This fiction was maintained throughout the Desert Campaign so that Rommel thought he was up against a stronger force than was the actual case. Wavell did this to good effect, first on the Italians and later the Germans. One of Wavell's early projects was the creation of the well-known Long Range Desert Group to harass the Italians and later to inflict much damage deep behind the lines of the *Afrika Korps*. This disruptive raiding technique was developed so that their enemy were often led to expect threats where there were none and not expect imminent attacks which could come upon them out of a clear blue sky.

The deceptive military initiatives grew as Wavell recruited a group of like-minded officers into a unit that he called 'A' Force under the inspired leadership of Colonel Dudley Clark. The colonel had created the first Commando and SAS units and his forces were the inspiration for similar units in the US Army, such as the Rangers. It was the purpose of A Force to create deceptive illusions for a largely unsuspecting enemy to their great discomfort and uncertainty, all of which Dudley Clark did with enthusiasm. The A Force operation used this unusual military technique across the whole of the Mediterranean region. One major achievement was to convince Hitler that the invasion of Greece by the Allies was imminent. This deception kept a large number of German troops and aircraft tied down awaiting an invasion that never came. To do this it was necessary for Dudley Clark to use intelligence to understand the concerns that Hitler already had about the danger of invaders landing on Greek beaches.

The A Force ensured that the Italian general opposing Wavell continued to be under the illusion that British troops were greater in number than his own. Not only did the deception make the Italians think they were facing a more sizable British Army, but it created the same illusion in the mind of the German Abwehr and this stayed with them for the rest of the North African campaign. Illusions helped greatly in Operation Compass, during which Allied forces counter-attacked Graziani's troop formation entrenched in Sidi Barrani. Ultra was still being developed at this time but the decrypts of Italian Air Force codes gave information on the movements and order of battle of the Italian Army. Operation Compass was originally intended as a five-day raid, commencing on 9 December 1940, and consisted of the relatively smaller British and Commonwealth force commanded by General Richard O'Connor. The Italians found themselves outmanoeuvred and soundly beaten by the following February, due in large part to the careless way they used their radios.

Wavell had all the information he needed from Italian radio traffic, according to Wilhelm Flicke, whose listening station was monitoring the region. O'Connor's bleary eyed but jubilant soldiers out in the desert had advanced 500 miles to take 130,000 prisoners including seven generals at the cost of little more than a thousand casualties. They were poised to advance on Tripoli and drive the Italians out of Africa when a decision in London robbed them of a great victory. Wavell was ordered by Churchill to send three divisions of his small force to Greece to assist the Greek Army to defend itself against the German invasion as a political gesture. Meanwhile, in Berlin Hitler was furious at the Italian defeat and had to send his troops to support the army of his main ally for the second time. General Erwin Rommel received his orders to take the *Afrika Korps* into the desert and make history.

The *Afrika Korps*

By mid-March the Italians had been thrown back to the borders of Tripolitania when the *Deutsches Afrika Korps* (DAK) arrived in the desert with Rommel in command of all German and Italian forces in North Africa. By 1941 a series of tank battles were being fought in the region with new rules of mobility and firepower that Rommel exploited with an expertise that often baffled the British. Panzer units would often appear unexpectedly out of the vast desert as the British lined their tanks abreast in the manner of a cavalry charge. They would press forward in wide vulnerable formation while Rommel's tactic was to concentrate his armour into spearhead formations and punch holes into a line of defence or an enemy armoured formation. His troops were organised differently as well, using a Mission Control, or *auftragstaktik* briefing, that gave detailed orders to a commander who would then give his men a clear objective and let them decide on the best way to achieve it. The method gave German soldiers the crucial advantage of not having to wait for orders when a situation changed as the British soldier often did. German soldiers gained a reputation for speed and flexibility in action which was particularly apparent in counter-attacks. There was another advantage that Rommel had which enabled him to act with the surprising speed and cunning of a Desert Fox of which the British were not aware.

In March 1941 the *Afrika Korps* was given an intercept platoon and then as a reward for success it was enlarged to form 621st Radio Intercept Company, equipped with receivers and direction finders modified to cope with tropical conditions. The new company was commanded by Hauptmann (Captain) Seebohm who seems to have been a gifted intelligence officer. His intercept evaluations carrying intelligence about British Army order of battle, strengths and intentions for Rommel to read were probably more detailed than the Ultra reports. British commanders had the impression that the German general had second sight in the way he fought his battles and forestalled most of the moves they made. In the House of Commons Prime Minister Churchill said, 'we are facing a bold and clever foe and I may say a great General', and later he acknowledged that the Libyan campaign had not gone as expected. Both sides had knowledge and intelligence of the other's order of battle and objectives but generally German intercepts and intelligence were probably more detailed and certainly more immediate than the British. The reason being their use of the 'Black Code' of the United States, which had been stolen by the Italians from the US Embassy in Rome just before America entered the war.

The Americans were watching the battle with great interest and had sent Colonel Bonner Frank Fellers as their US military attaché to British Headquarters in Cairo. He was regarded as representing a friendly nation and ally by the British, even though he himself was an Anglophobe. Fellers was given carte blanche to tour all units of their army so he got to know the strengths and weaknesses of the British dispositions and wrote detailed reports of men and equipment, their positions and even future objectives. He faithfully sent these back to Washington on a regular basis but was unaware that Rommel, the Desert Fox, was listening to those radio transmissions as he had the Black Code Book and Hauptmann Seebohm to decipher it for him. Wilhelm Flicke at the Lauf listening station was also monitoring Fellers' transmissions; he describes in his papers how Fellers' reports detailed how the British planned to drop parachutists near Rommel's main airfields and blow them up. Fellers sent a report about these raids at 8 a.m. and it was deciphered and on the desk of the Führer by 10.30 a.m. It was then confirmed to Rommel who had a day to warn his airfields to expect the raids. The actions were, of course, a disaster for the British with the exception of one airfield where they ignored the warning. Rommel was indeed an aggressive and capable commander of the *Afrika Korps* but the signals intelligence that he was given made him doubly effective; his luck however, was beginning to run out. The supply of excellent intelligence that he was getting would soon dry up but there are two versions of the way that the British were alerted to German interception of Colonel Fellers' reports to Washington.

Rommel used Seebohm's company to listen closely to what the British and Commonwealth troops were saying as they planned attacks on German positions but more importantly his listening company tuned into the almost daily reports by Fellers to Washington. It was easy to identify his transmissions as they always started with the MILID WASH or AGWAR WASH call-signs. To get the best reception Seebohm had to get as close as possible to the British operators in the frontline. This gave General Auchinleck, who had replaced Wavell, the opportunity to target the German listening company so on the night of 9 July 1942 the Australian 2/24th Battalion stealthily approached the listening company's camp. On the frontline at Tel el Eisa they crept past an Italian battalion that should have been providing a guard for the German listening unit but who were sleeping. They surprised the Germans and after a short, sharp engagement the signals intelligence unit was 'put in the bag' with much captured equipment, papers and prisoners. Hauptmann Seebohm was badly wounded in the attack and died later in a Cairo hospital. This was probably the first operational listening unit that the British Army had captured complete with documents and equipment, and it gave them an insight into the working methods of German intercept units. It was of great use in correcting many mistakes that British signals operators were making. The captured documents would almost certainly have contained Colonel Fellers' transmission intercepts and, of course, the American Black Code Book. Rommel's reputation as the Desert Fox was based on his prescience in out-thinking the British commanders in the field, but the capture of his listening company coincided with the end of his run of victories. Would the German general have achieved the reputation he had without the constant special intelligence that he received? A question that the Fellers security failure raises is how many British lives did it cost in the desert campaign and if the contribution of Bletchley Park is said to have shortened the war by an estimated two years how much did Fellers lengthen it by? Colonel Fellers was relieved of his post in July 1942 and sent home where the American government awarded him the Distinguished Service Medal for 'reports of clarity and accuracy'.

The Australian raid is one probable source of information causing Fellers' security breach, but Wilhelm Flicke's papers reveal another. He recounts how he was at home on 27 June 1942 having his breakfast and listening idly to the *Deutschlandsender* radio station. It began with a programme entitled 'Events in North Africa' which included a discussion on political and military matters.

The account did not make it clear if the programme was a play or a discussion. One of the characters took the part of an American military attaché in Cairo and described how he was disclosing extensive Allied information to Washington and was intercepted by a German Army listening post. The drama played too close to the situation that Flicke was monitoring in real life from his listening post on a daily (and sometimes even hourly) basis. He was stunned as he listened to a major intelligence secret being discussed over the radio and he wondered what would be the outcome. Some thirty-six hours after the broadcast went out the messages from Fellers ceased and, although Rommel's replacement intercept unit searched the wavebands to try and intercept the MILID or AGWAR transmissions again, nothing was found. The author tried to confirm this story but the *Deutschlandsender* archives had been destroyed by bombing. If true, it seems to constitute the most public disclosure of highly secret information on record. Those are the two versions – take your pick.

The war at sea extended into the Mediterranean with German attacks on convoys to supply Malta and the British sinking German ships, in particular oil tankers trying to supply Rommel's *Afrika Korps*. Ultra's guidance to a ship's movements could be traced from an Italian port and sometimes it was even possible to identify the cargo. The result was that Rommel was short of everything but particularly oil, the shortage of which had begun to affect his ability to manoeuvre his forces, particularly his tanks. In the summer of 1942, a U-boat in the Mediterranean captured a British ship along with the radio codes used by the British armed services in the Mediterranean. For a few weeks the Germans were able to read all British communications traffic until suspicions arose and new code books were distributed to all radio operators in the region. This was one of the last successes German signals intelligence had in the Mediterranean region as Bletchley Park increased its ability to intercept and break coded transmissions using lessons learned from Rommel's captured intercept unit. As the British and American forces closed in on the *Afrika Korps* in Tunisia, preparations were made for a last stand and impending capture. The communications intelligence officer in command of the intelligence company replacing Hauptmann Seebohm's unit asked to be evacuated to serve in Italy. The order of Hitler, however, was clear – no men or equipment would be allowed to leave Africa thus another of the German Army's dwindling number of intercept companies vanished. Their equipment and papers had been destroyed by the time that the Allies caught up with them but from about this time the advantage in the signals intelligence battle which had see-sawed from one side to the other slipped out of German hands. From now the Allies controlled the signals intelligence field increasingly and so it remained for the rest of the war, a reality unknown to the German High Command until after the war's end.

The Torch Landings

As Rommel retreated from El Alamein, both he and the OKW in Berlin were taken by surprise as a huge assault of Allied troops landed on the coast of Morocco and Algeria in Operation Torch – the Allied invasion of French North Africa. The Torch fleet consisted of 1,500 ships carrying 90,000 men and later another 200,000 with all their equipment and supplies, including ammunition, transport and tanks in what was the greatest invasion fleet known up to that time. It was also one of the great deceptions of the war and its unannounced arrival was a heavy blow to German intelligence as Admiral Canaris, the Director of the Abwehr, was unable to give an apoplectic Hitler any evaluation of the battle order of this Armada. In fact the Abwehr had been subject to a double bluff using many forms of misinformation mainly by signals traffic. It had not escaped the attention of the Abwehr's network of agents and informers that a major fleet of ships was being assembled for an operation – the question was what operation. A deception was mounted by Committee 20, which when expressed in Roman numerals (XX) aptly became the Double Cross Committee, who were to protect the great fleet of ships. The Committee had become a past master at military deception, with one of its main achievements being the identification and, on occasions, 'turning' of German spies that entered Britain. Using the agents they now controlled, the Committee were able to manipulate German espionage efforts to a considerable degree. They led the German High Command to think that the objective would be Greece or Sicily or even Italy, but the most favoured one was the invasion of Africa using Dakar as a staging area for further incursions. The intelligence arm of the Kriegsmarine, however, was more on the ball than the Abwehr. It had assembled a fleet of forty U-boats in a line from Gibraltar to Dakar to await the convoys of merchant ships as they appeared over the horizon.

The British Secret Service used their counter-intelligence agents and other signals deceptions to send out hints and clues about a great convoy of troop ships northward-bound from Sierra Leone. This was convoy SL 125 that consisted in reality of a number of empty cargo ships acting as bait for the German submarine flotilla. The Kriegsmarine took the bait and immediately ordered their U-boats to leave their station and fight a running sea battle with the decoy convoy off the East African coast sinking thirteen ships in the convoy. Meanwhile, the task force slipped through the Gibraltar straits and on to the shores of North Africa with the loss of just one transport. American forces landed on 8 November 1942 and expected to be welcomed with open arms by the French. The Americans had always enjoyed good relations with them but were surprised at the brisk resistance from French marines in Algiers and Oran, although there was almost no resistance in Casablanca. The French commanders on post in French North Africa had been appointed by Marshal Pétain, head of the Vichy government, and were uncertain how to receive the US Army. A few days after the Allies landed in North Africa, Hitler ordered the German Army to occupy the rest of France not conquered in 1940. That act released French commanders from their commitment to the administration in Vichy so a ceasefire was ordered all along the Algerian coast. Operation Torch was the Abwehr's greatest intelligence failure so far and left Hitler with a situation he was not able to counter. Indeed, Torch was going to contribute to one of his greatest defeats.

The *Afrika Korps* was now held in a pincer movement between the British Eighth Army on one side and the Anglo-American forces of Torch on the other. With its back to the sea, the *Afrika Korps* was trapped but not tamed. It struck hard at the new, untried and over-confident American troops at the Kasserine Pass, Tunisia and hoped to split them from the British forces but the Torch forces slowly recovered their ground. Rommel then turned on Montgomery's Eighth Army which proved a very different battle-hardened force, and lost many of his tanks in this attack. It was the beginning of the end for the *Afrika Korps* as the Long Range Desert Group found a way through the mountains for Montgomery to turn the flank of the Mareth Line, Rommel's last strong defence.

The desert campaign came to an end with the surrender of a quarter of a million men and their equipment, more than had been captured at Stalingrad. The 'soft underbelly of the Axis', as Churchill called the southern shores of Europe, was not going to prove so soft for the Allies. The Desert Campaign had been a proving ground for co-operation and military effectiveness for the Anglo-Americans. In addition, the British had at last developed an army that could defeat the Germans on their own terms and learnt valuable lessons along the way. Ground-to-air co-operation was essential, but the most important lesson was the almost cultural development in the army of the acceptance of military intelligence as a weapon. By this point, aspects of military intelligence had become accepted by senior officers in both the British Army and US Army as a guide to action against the enemy.

How many lives lost
N° U Boats sunk?

10

The Second World War – The Middle

Russia

The High Command of the Wehrmacht (OKW) were not good at accepting intelligence evaluations that they disagreed with. Nowhere was this more obvious than during the Russian campaign. The papers of Colonel Randeweg, who was commanding the German intercept units in southern Russia before Operation Barbarossa in June 1941, showed a clear picture of the Red Army and Air Force order of battle. Intelligence from the Abwehr reported the Soviet Air Force would be able to field 10,000 aircraft in the event of war and also that the Russian aviation industry was capable of a high production of planes. The Luftwaffe's general staff discounted this figure and decided that the number of operational machines was in the order of 3,000 and losses would not be easily replaced. They were much encouraged when a captain of the Soviet Air Force was captured and gave up the Russian key to the air code so Luftwaffe fighter squadrons were able to shoot down over 100 Russian aeroplanes in two weeks in air battles over Minsk. The code was changed within a couple of weeks but the damage to the Soviet Air Force was done. Within a month of launching Barbarossa, more than 3,000 aircraft had been shot down, and yet they still were able to show a strong presence in the air. The figures were, therefore, distrusted by the High Command who decided to do an audit and count the number of crashed aircraft. They came to the surprising conclusion that the claim of 3,000 shot down was a considerable underestimate. A revision was then made by the High Command in July, which enabled the OKW to claim that 6,233 Russian aircraft had been destroyed. This was an indication of the attitude of the German High Command to the often incisive intelligence that the Abwehr was able to gather.

The intelligence community was regarded by Hitler as being anti-Nazi and, although all would profess to be German patriots, he was right to some extent, which put them in a difficult and sometimes dangerous position. Chief among the suspects was Admiral Wilhelm Canaris, the Director of the Abwehr and spymaster to the German Wehrmacht, and among others was Wilhelm Flicke, senior cryptologist in German signals intelligence. Flicke's constant references in his papers to the glaringly obvious shortcomings of the German High Command in its direction of the war were almost treasonable. The almost perverse and dismissive misreading of intelligence evaluations created a situation which made not only his, but also his colleagues' attitudes clear. The unease of the German public about the invasion of Russia was palpable, although it was not evident in the German media which was strongly controlled by the Ministry of Propaganda. The advance of 3 million men of the Wehrmacht, almost half of Hitler's armed forces' strength, into the Russian hinterland started with the support of her allies. There were eighteen divisions of Finns (they had a score to settle), sixteen divisions of Romanians, three of Italians and another three of Slovaks, as well as a scattering of Croatians and Hungarians. This vast juggernaut of men and machines was launched against a Red Army whose officers were well aware of the German threat. Stalin refused to recognise it, however, even after Churchill, with the aid of Ultra, warned him of the preparations for Operation Barbarossa.

Almost 200 divisions of the Red Army that were stationed near to their borders were thrown back or overwhelmed in the first onslaught of the attack. Soviet signal communications were thrown into chaos as they reeled before the assault, causing the security of radio messages to become so lax that the German Army intercept units were able to get a clear reading of the order of battle of their opponents. By the end of September, the Red Army communications system began to improve as the army got over the initial shock, although they had lost many experienced radio operators in the first offensive. The Russians had also lost a vast quantity of equipment, but it was mostly out of date and about to be replaced. As war began, Russian industry set about enacting the miracle of production that would churn out enough supplies and equipment to enable the Red Army to later turn the tide in their favour. Also, as the shock of the invasion by the Panzers was absorbed, security disciplines began to return to Soviet radio communications. It was not so in German signals communications, however, as the demand for radio operators and particularly evaluators of the intelligence data had increased by a factor of ten. Experienced personnel were spread more and more thinly to satisfy a rapidly expanding intercept service and this led to fatal mistakes in evaluations sent to the OKW. A communiqué was issued saying that no unified command structure of the Red Army could now be recognised, implying that it was breaking down into isolated groups. The reverse was true; the Russian front was strengthening, but the symptoms were not being recognised by the newly recruited cryptographers in German intercept stations. Their stations intercepted what they thought was the Red Army's chaos in radio communications, but it was actually resorting to the norm. When security practices returned to Russian operations, fewer transmissions were received and from fewer locations, hence the conclusion was drawn that the organisation of the army was falling to pieces. It was not. Hitler always assumed in his crusade against Bolshevism that the people and the army's resistance would collapse as anti-Communist forces came to their aid. This was not true as the Russian people saw the German Army as the invaders that they were, resisted very strongly and began to fight what was called the Patriotic War.

It was assumed that the resistance of the Russian soldiers was being maintained by political commissars who were behind them with weapons to keep the men in place in the frontline. As a result, intelligence reports that the Red Army was crumbling were welcomed and in a speech to the German nation he said, 'The enemy is broken and will never rise again'. This was followed by a statement that:

We have been so forehanded that in this mighty war of materiel I can now cut back production in many lines because I know there is no longer any opponent whom we cannot overcome with the stock of ammunition on hand.

Production lines were substantially cut and the supplies to the front were later greatly curtailed; as a result, German soldiers suffered severely from lack of ammunition on the Russian front. This was self-deception on a grand scale, but there was worse to come as the German media reported that the Russian Marshals Voroshilov, Timoshenko and Budyonny had been relieved of their commands and turned over to the GRU (Soviet military intelligence) in a brutal purge. The German newspaper headlines shouted, 'They are Silent in Moscow' – and so they were. The three experienced commanders had, in fact, gone beyond the Urals to train the many new divisions released from duty on the Japanese frontier.

Wilhelm Flicke's intercept station at Lauf had, meanwhile, been intercepting much radio traffic from the region east of Moscow and come to some disturbing conclusions. In October Flicke's superior, Colonel Kettler, reported forty new Red Army formations of the size and nature of a division being trained and put into an army reserve of formidable size. Kettler was able to report the nature of each division, such as tank or infantry units, either motorised or cavalry, their composition and strength in men and machines, their equipment and ammunition states and command structure. Flicke then helped him compose a report of his findings in great detail to send to his commander, General E. Fellgiebel. Flicke then added in his papers that this general was later murdered by Hitler following the bomb plot in July 1944. The report was then sent to the OKW for the attention of Hitler where its findings were met with an immediate rejection. A note to Fellgiebel on its cover from Field Marshal Jodl said that Kettler should be put out of business. It was countersigned by SS Superior Group Leader Fegelein who noted that this was also the opinion of the Führer. Fegelein was shot on Hitler's order just before the end of the war. By November there was increasingly stiff resistance from the Russians in front of Moscow as reinforcements to the Russian line began to arrive in total radio silence, a great achievement in radio operations given a movement of that size. A stirring moment in the war is brought to life in the newsreels of Stalin taking the salute at a parade in Red Square in Moscow as battalions of Russian soldiers marched past him and straight on in to the trenches just outside Moscow to hold back the German Army.

German Army commanders eventually began to see the light and General Halder, the Chief of Staff, wrote in his diary that it was becoming ever clearer that they had underestimated the Russian colossus. The High Command had not wanted to get bogged down in a positional war of entrenchment. They had expected 200 enemy divisions to oppose them but so far they had counted 350 and, although they were not always properly equipped or led, they were always there. Whenever a dozen of these divisions were destroyed or captured, another dozen would immediately replace them. In addition, the Germans' long line of supplies was being increasingly disrupted by Russian partisans while the Red Army were close to its source of supply. The Germans found themselves in front of Moscow with little or no winter clothing in -30°C: their skin froze on to the metal of their guns as they touched them and tore off their hands if they tried to pull them away; the oil froze and the tanks could not function or the guns fire; and the radio sets would not work as the batteries froze, so the German generals had the dilemma of withdrawing and leaving their equipment or staying put to be overrun as their weapons did not work. Then to the amazement of the Germans, the Russians attacked them in early December with forty divisions of fresh, well-equipped reinforcements from Siberia, who considered the Moscow weather to be relatively warm. They fell upon the German troops like wolves and the whole of the German Army Centre Group retired in disarray to a line 90 miles distant from Moscow.

! see PREVIOUS PAGE.

German intelligence had failed Hitler once again after a complete lack of warning of the Torch landings and now an underestimation of the strength of the Red Army by almost half as much again. Russian production achievements had massively exceeded the Abwehr estimate and their T-34 tanks were beginning to appear on the battlefield in their thousands.

American aid was also coming to their new allies through the Middle East. Hitler berated Canaris for the failure of his organisation in public and the Admiral knew that he would pay a high price for the failure in a way that would only help Himmler's SS who were building a separate intelligence organisation of their own. The intelligence service, or rather services, were in conflict and the results would be disastrous, but the Abwehr knew it had to pull off a coup to get Hitler's approval. Meanwhile, in Russia things were going from bad to worse as the German Army reached its high water mark in their advances into Russia and then the tide began to recede.

In September 1942 the long crucifixion of the German Sixth Army at Stalingrad began as they advanced to the banks of the Volga. By November the Russians' pincer movement had been launched and the two arms met at Kalach behind the German Army. Russian troops encircled the Sixth Army, although the Luftwaffe was able to lift 50,000 men out of the trap. A final surrender in February of the following year had cost the German Army almost a quarter of a million men as casualties or prisoners. This great public humiliation for Hitler and his army convinced the world that the Wehrmacht was not invincible. The last message from General Paulus, commanding the doomed Sixth Army, was intercepted by the Lauf listening station and Wilhelm Flicke recounted how the message brought in by a duty officer read, 'My Führer, in future follow more the advice of your Generals!' The document was passed from one adjutant to another until it landed on the desk of Field Marshal Keitel. He ordered that it be taken into the office of the chief of the supreme command of the armed forces of the Third Reich immediately. The author himself knows how impressive and overbearing the entrance to Hitler's office was, as he visited the bombed and ruined Reich Chancellery just after the war. The double doors were 30ft and each was 6ft wide with heavy sculptured bronze images depicting German myths of the past. None of those brave soldiers in attendance on the Führer ventured to volunteer to pass through those doors to take the document to him – Nazi Germany occasionally copied the Greek practice and killed the messenger bearing bad news. Finally the paper was placed in a portfolio with others and laid on Hitler's desk in a casual manner, and then a small but apprehensive group waited outside.

The silence did not last long, from behind the massive double doors came the sound of breaking vases and chairs being overturned. Hitler tore the telegram into smaller and smaller pieces and ranted and raged at everything and everybody but himself. Keitel was summoned and listened to the strident demands for the degrading of Paulus from his rank of field marshal, confirmed for him the day before. It was too late as the announcement had already gone to the press so Keitel stood, stony faced and took the tirade of bitterness and frustration. In the ruins of Stalingrad the new Field Marshal Paulus marched what was left of his army through the snow into a long captivity from which only 5,000 would return. Flicke made a surprising assertion in his papers about another battle at Kharkov further south, which was regarded as a masterpiece of manoeuvre by one of Hitler's best generals, von Manstein. He persuaded the Führer that a positional war such as Hitler had experienced in the First World War needed to give way to one of movement. The general was allowed to manoeuvre the Red Army into a trap as exhausted Russian troops ploughed on to the extreme of their supply chain before the German Panzers struck. The world saw von Manstein's victory as a model of defensive mechanised warfare but Wilhelm Flicke, whose listening station measured the action blow by blow, did not.

Flicke compared Kharkov with Napoleon's 'victory' at the Battle of Borodino where the Imperial Russian Army was defeated but Napoleon's army was so damaged in the battle that its ultimate fate was decided. Flicke thought that the way that German forces were weakened at Kharkov enabled the Red Army to prevent the Germans from taking Stalingrad or the oil fields in the Caucasus. The real decision in Russia, Flicke asserts, did not take place at Stalingrad but earlier at Kharkov when the Red Army inflicted such severe losses on the Germans that the timetable of the depleted German divisions was upset as they moved on to the Kuban area, the Caucasus and the Volga. The Russians sustained terrible losses in men and materiel but were able to replace them, while the Germans not only lost men but also time.

Soviet espionage activities both within German-occupied Russia and also in the German homeland were well known to the Abwehr. Flicke tells how detailed German operational plans for the 1942 spring offensive were known to the Russians as was the plan to advance across the Volga and into the oil fields of the Caucasus. Areas of assembly of German Army formations designated to carry out these operations and their order of battle, with the numerical strength of the army and their allies, were all known to the Russians in detail. Units from battalion size upwards had been identified with even the names of unit commanders and the numbers of tanks, guns and planes available and those not available but under repair for the coming campaigns. The attack plans for the coming summer offensive of the three German Army groups in Russia had been established and the *Rote Kapelle*, or Red Orchestra (the Soviet spy ring), had reported them to their Moscow control. In short, the Russian Intelligence services knew as much about the German Army and its order of battle and future plans as Bletchley Park did, although probably by different means.

Intelligence services of all beligerent countries had planted networks of radio agents around Europe, notably the Russians. From the 1930s she was the first country to develop espionage networks with an international dimension. The Soviet Union gradually set up networks of shortwave radios reporting to a control station in Moscow and other lesser control stations in other European countries. Wilhelm Flicke's intercept service recognised and reported these stations to the OKW who assumed they were a propaganda network for Communist International purposes aimed at spreading the word of Communism across Europe. There was little indication of the network's size as most of the stations kept a discreet radio silence, nor any indication of their nature as those transmissions that were received were in a code that remained unbroken and only made infrequent transmissions. The network was, in fact, the radio communication system for the Russian intelligence service designed to collect and transmit intelligence gathered by Communist sympathisers across Europe acting as agents. There were many such people before the war who were told to observe and report on all aspects of armed forces activity in their country. In addition they were ordered to report on industrial strengths and weaknesses of the economy in European countries, particularly Germany. Agents were expected to assess the production capabilities of their country and its technology, including the searching of patents. Members of the network were also tasked with monitoring political events and identifying politicians that might have an effect on the wellbeing of the Soviet Union. Soon after the outbreak of war in 1941, the German state security police estimated that there were 120,000 agents and fellow travellers serving the network, who all kept a low profile as they reported their findings through embassies. The long-term investment in establishing this huge intelligence operation was about to pay off in a big way.

The number of shortwave stations that sprang up in all territories occupied by Germany as she declared war on Russia in 1941 increased enormously. Wilhelm Flicke and his colleagues named the network the WNA net, which was taken from the call-sign of the Moscow station directing it. Dozens of radio transmissions on many shortwave frequency bands suddenly came to life and connected to what was to prove to be the largest espionage networks in Europe. The German listening service was overwhelmed; Lauf and other listening stations counted over 600 radiograms in the month of August following the invasion of Russia. Transmissions could be heard coming from every European country but most disturbingly some from within Germany itself. The building of a clandestine intelligence operation of this size and scope must count as being among the most successful the world has ever seen.

The most powerful network in this Russian intelligence assault on the Third Reich's secrets was undoubtedly the Red Orchestra (*Rote Kapelle)*, as the Abwehr named one of its Soviet espionage networks. The translation of the word *kapelle* into English is uncertain as it has a double meaning; it can either be a chapel of religious worship or alternatively an orchestra. The author favours the orchestra term because Flicke refers to the operators in the network as musicians in his papers and the director in Moscow as the conductor. The Soviet espionage organisation had three largely unconnected parts each of which gave the Abwehr and Gestapo many headaches. These were the Schulze-Boysen group, the Trepper group and the Red Three (*Rote Drie*) network, with a base in Switzerland. The first of these was brilliantly run by two men, Harro Schulze-Boysen and Arvid Harnack, who were both Communist sympathisers who managed a disparate group of over a hundred anti-Nazi agents, most of whom had been members of the Communist Party in Germany until forcibly dissolved by the Gestapo. Many in this motley crew did not involve themselves in espionage directly but observed and reported matters of interest to the network leaders. An inner circle of more active agents used their surprisingly good skills and contacts to enhance the observations of the others. Horst Heilmann worked with the Wehrmacht on decoding signals; Johann Gradenz sold aircraft spares to the Luftwaffe and knew about aircraft production; and Herbert Gollnow was a policeman and had access to counter-espionage secrets. Other shadowy figures worked in the German Foreign Office, the Ministry of Labour, and the Berlin Council, holding mainly government positions. From this mixture of observers and activists the Schulze-Boysen Group were able to accumulate a surprisingly rich vein of intelligence evaluations to report to its Moscow control.

The Abwehr's painstaking decoding of the texts left them astonished at the high quality of the intelligence in the reports, which might concern anything from the movement of thirty army divisions being transferred from west to east, or 400,000 German soldiers holding strategic points in Italy to guarantee that her government would not make a separate peace. At another level, a technical description of a new anti-aircraft gun could be included, as well as more internal political matters. For instance, Hitler's willingness for the Finns to make a separate peace with Russia once the Germans occupied Leningrad for the purpose of shortening her line of defence and to enable her troops to be more easily supplied was one item reported on. Details of the German war machine and its manufacturing base was another; a breakdown of the statistics of Luftwaffe strength in the air and how the 22,000 machines of first and second line aircraft were deployed. Losses of planes were also enumerated, such as the fact that ten to twelve dive bombers were being built a day but forty-five planes had been lost on the Eastern Front from 22 June to the end of September. These and many other aspects of military and economic information emerged as the horrified German cryptanalysts worked to clarify the contents.

The case caused German intelligence great concern. After breaking up the Schulze-Boysen group, the Germans found that dedicated Communist agents had been planted many years before the war without arousing any suspicion. They had used their positions in industry to gradually make excellent connections and become trusted by leaders in industry and the armed forces. The penetration and breaking down of the network in Germany by the Abwehr was a blow to Soviet intelligence, particularly as the arrests led to the discovery of networks in France and other occupied countries. The link to Moscow using the Schulze-Boysen Group's captured radio sets was kept up, although Moscow soon realised that their network had been blown. The Russians kept up the double bluff of pretending that the network was still working, as they wanted to distract the Germans while the Soviets strengthened and built another intelligence network.

This new network was the Trepper Group, a Soviet espionage ring run by staunch Communist Leopold Trepper who posed initially as a Canadian industrialist. He started to trade in clothes and underwear in Brussels as a cover for his espionage activities and developed business interests in France, Belgium and Germany before the Second World War. He created intelligence networks of Communist agents in those places using his business activities as a front for his agents while selling black market goods to German forces occupying those countries. He supplied Hitler's Organisation Todt (the Third Reich's civil and military engineering group) with materials and to do this he changed his persona to that of a German businessman. He used social occasions and dinner parties to cultivate high-ranking German officials to elicit information about troop movements and building defence projects. In late December 1941 his transmitter in Brussels was detected by a directional indicator and was promptly shut down by the Abwehr and, after a long chase, he was arrested in Brussels. After interrogation, he agreed to work for the Germans by transmitting disinformation to Moscow, although he managed to include hidden signs to his controller giving warning of his plight. In September 1943 he escaped and went into hiding with the French Resistance, but by then all the members of his groups had been arrested, including the well-known French agent Suzanne Spaak who was executed at Fresnes Prison just two weeks before Paris was liberated. Trepper survived and, after the war, returned to his old business of clothes wholesaling and quite possibly continuing espionage for the Soviet Union.

Late in 1941 the intercept services at Lauf began picking up transmissions from three new operational stations, one of which transmitted from Switzerland. Their transmissions to Moscow soon established them as agent stations in the Soviet Red Orchestra. The Abwehr christened this new station *Rote Drei* or Red Three in the network. It took the German cryptographers until well into 1944 before they could decrypt any of the enormous volume of traffic that passed on the shortwave links to Moscow. The Abwehr rated this network as especially dangerous as it operated in Geneva in neutral Switzerland, outside the security sphere of the Third Reich, although that did not stop them investigating the operation. Its first question was easily answered as they identified Alexander Rado, a Hungarian national, as the main agent in the network. He lived at 22 Rue de Lausanne, next door to the Comintern International offices running a Communist propaganda programme around the world. Rado, whose code name was Dora, had two radio sets allocated to him and was supervised by his director from Moscow Central. Wilhelm Flicke says that one of those sets was at another address in Lausanne, 2 Chemin Longerai, where an Englishman A.A. Foote was living, but how Flicke knew any of this is not clear. He goes on to say that Foote's cover name was 'John' and he did all his own cipher work and worked independently of Rado. This is strange as the Russians and particularly those in the intelligence community had a deep distrust of the British.

Rado established the *Rote Drei* in 1937 with a small staff in Geneva to cope with the heavy workload of transmissions. German intelligence established the names of all of them and their code names but could get no further information. In the critical period of the battles for Stalingrad and the Caucasus the radios were never quiet and, following the encirclement and capture of a German Army at Stalingrad, the Russian intelligence service faced the most important problem in its history. After the Red Army's initial breakthrough of German lines, what was the enemy's situation? Did they have enough reserves to strike a counter blow? Did the Red Army risk falling into a trap as they advanced so rapidly or could they pursue the enemy safely on the Southern Front? This was the finest hour for *Rote Drei* and indeed the Soviet intelligence service; they rose to it magnificently by answering the Red Army's questions in detail. The often hourly reports detailing the order of battle of each of the three army groups of the Germans were received in Moscow and proved unfailingly accurate and timely. German intelligence operatives have told the author that the war was won in Switzerland, but how was *Rote Drei* getting its information and who were its informants? The German security agencies tried every means they could to find the leak that was haemorrhaging away the strength and dispositions of the German Army, but to no avail. They knew all the people in the *Rote Drei* office in Geneva and watched them closely, as they did with hundreds of people in the Führer's Headquarters, but could not find a hint of the leak of such critical and wide-ranging information. Wilhelm Flicke wrote that, as the tide turned against Germany in the late 1940s, 'The Rote Drei's source of information remains the most fateful secret of World War 2'.

It was obviously a dark mystery to the Germans, but there is a simple explanation; the mysterious A.A. Foote (John), either known or unknown to Rado, was a member of the British secret service, or MI6, whose life story could fill another book. Both he and Rado were being hounded by the Abwehr, together with the Swiss counter-espionage organisation BUPO, but all the time Foote was in touch either directly or indirectly with London. The critically important information that Rado, trusted by Moscow Central, was passing on was intelligence that came from Foote. The content was so detailed and accurate over a long and critical period that it must have come from Bletchley Park. The indirect use of *Rote Drei* had a double advantage to the Park: first, the information came to Moscow from one of their own tried and trusted people; secondly, it safeguarded the Park's security as Rado was seen by the Germans as a wizard at finding mysterious sources of information that they could not identify. Meanwhile, the Abwehr and Gestapo were mesmerised into strenuously seeking the answer in the Führer's Headquarters or the inner circle of the Wehrmacht and never thought to look further afield.

Russian Battlefield Intelligence

The tactical battlefield operations of German intercept units in the field against Russian troops had an entirely different nature to those on the home front described above. Colonel Randeweg, commanding the intercept detachment in the German Army Group South in Russia, recounted his experiences that were probably similar to every intercept unit in Army Groups Centre and North on the Russian front:

> The vastness of Russia's steppes, with little in the way of good roads and almost nothing resembling a commercial or military communications system, left the Russian army with no option but to use radio to contact its formations. German signals intelligence operations therefore concentrated on long-range interception operations to determine the battle order of the Soviet army and air force west of the Ural Mountains.

The mission of Randeweg's units was to establish the current radio techniques of Red Army operators and what German interceptors could find about their unit's command structures and strength. The scenario gained from these operations showed a picture of the Soviet Air Force that was very different to the evaluation accepted by OKW intelligence officers as has been seen earlier. The lack of information available to Army Group South about the Red Army caused the Germans to make a grave error in underestimating the Red Army's strength.

Russian military signals security in their frontline units was not good and tank units in particular gave themselves away by faulty security procedures before and during attacks which made German intercepts very effective. In particular, careless requests for fuel gave away their positions and condition and transmissions from tank commanders made them particularly vulnerable. In July 1942 the Russian 82nd Tank Brigade had been trapped in a large pocket by the German Ninth Army who intercepted a plain text message discussing a break-out. The Brigade Commander asked about the axes of movement for his formation and was advised on the best location for an escape. The general of the 9th ordered the escape route to be lined with his tank-killer 88mm guns which decimated the Russian T-34 tanks and prevented the break-out so the remnants of the brigade retired into swamp-land for cover. Soon messages were intercepted requesting assistance in towing their T-34s out of the muddy swamp so the German radio operators used a deception on tank commanders, pretending to be Russian and asking for their position so the towing vehicles could find them. They then used the co-ordinates to direct artillery fire on the position and, still pretending to be Russian, the operator was able to keep in touch with the tank commander until his tank was knocked out and went off the air. Finally the Russian divisional staff tried to rescue some of the brigade from its disaster by trying to reorganise the troops and ordering them by radio to assemble at designated points, which then came under further intense artillery fire. The effect of the bombardment could then be checked by transmissions from surviving Russian operators asking for help. The whole of the Russian 82nd Tank Brigade had been wiped out due to lack of radio security. Frontline Russian operators would find it difficult to equal this kind of devastating efficiency shown by German intercept companies.

In the autumn of 1943 German forces were encircled near Cherkassy in a similar way to that of the 82nd Tank Brigade and successfully broke out by intercepting Russian signals in the operation. At the behest of the German propaganda machine, the commander of the brigade of tanks told how he had been able to direct movements of his armour by use of interception of Russian transmissions. An account of it was printed in the German press and after the Russian frontline security of messages improved.

Another aspect of action in Russia was the use the Russians made of partisans. While serving as commissar with Budyonny's cavalry, Stalin's role as a partisan in the 1920s allowed him to observe the defects of the Russian radio services, and it taught him much about the value of disrupting the enemy's supply lines. A band of guerrillas in the vast empty Russian steppes could attack the lightly defended supply routes of the Germans whose first intimation of an attack would be the swish of the skis of the attacker before the destruction of supplies or ammunition intended for frontline units. A major partisan offensive was launched by Stalin using such guerrilla tactics to disrupt the German Army's ever-lengthening supply lines, causing a major headache for their listening posts. Each band carried a shortwave radio and took orders from their parent Red Army formation headquarters as to where to strike and when. Listening posts allocated to identifying these radio sets complained regularly that it was impossible for them to keep track of literally hundreds of tiny mobile stations transmitting in their sector.

As the tide of battle turned from the high water mark of Stalingrad and then Kursk, the Russian use of radio communications became more adept and the German interception service was gradually overwhelmed. As the Red Army fought along its savage 3,000-mile journey from Moscow across Russia and Poland and then through Germany, increasing numbers of German intercept companies and their equipment were destroyed or abandoned. The Russians, on the other hand, became more and more confident in their radio communication service and took less and less care as they advanced. By the time the Russians were on the outskirts of Berlin they were confident enough not to encode messages. The Germans, on the other hand, were using their skilled intercept personnel as infantry in a last ditch effort to hold back Russia's avenging army. In the last of his papers, Wilhelm Flicke asks the great why. Why, when the intercept service revealed the growing strength of the enemy on all fields, did they not stop the war when they knew it could not be won? German soldiers continued to fight as the Russians advanced into the gardens of the Reich Chancellery and the doors of Hitler's bunker in Berlin. The most convincing answer that I received came from a German ex-soldier, wounded in battle with the loss of an arm: 'It has to do with the German culture,' he said, 'we continued to fight because we were never told to stop.'

Japan

Japanese forces attacked American forces in 1941 and became an active part of the Axis alliance. Japan had a tradition in cryptographic matters dating back to the seventeenth century but their intelligence service proved quite inadequate to the task in the opening stages of the war. Their intercept service in the inter-war period was divided between the Tokko, or civil section, for the foreign ministries and so on, and the Kempei for all military purposes. The Japanese Army cryptographic service used Polish advisers in the 1920s while her navy received signals instruction from the Royal Navy well into the 1930s. The onset of war ensured that military needs took the lead in signals intelligence techniques and resources, with some help from Italian and German cryptographers. America's inability to decipher Japanese code messages much before 1943 contributed largely to her lack of understanding of Japanese foreign policy objectives leading to the war. During this critical period, it probably also contributed to the lack of appreciation of the effect the Russo-Japanese neutrality treaty would have on Japanese military objectives. The treaty with Japan contained assurances that Russia would not join in the conflict in the Pacific and attack Japan. This was a deciding factor in Prime Minister Tojo's decision to attack America and Britain in the Pacific. The Germans strongly held that discussions should have taken place between the Axis partners before such a treaty was signed, which was made clear to the Japanese ambassador Kurusu in Berlin in a furious dressing down at the German Foreign Office. When America's penetration of Japanese codes began, it played a part in many actions in the Pacific, particularly the Battle of Midway. Japan's action in going to war triggered by the Russo-Japanese treaty has been estimated as lengthening the global conflict by about a year and a blow to the British Empire in South East Asia.

Japan was less effective than Germany and even Italy in using signals intelligence against its enemy; they had been fighting in China and Manchuria against a less sophisticated army than their own since the early 1930s. They had not needed to measurably improve their intercept services during that time, particularly on the subject of transmission security. Germany blew hot and cold on the subject of co-operation with her Japanese ally on the field of intelligence and particularly signals intelligence. A key figure in talks was Hiroshi Oshima, the Japanese ambassador in Berlin, a sociable diplomat with a military background who was courted by all the government circles, in particular Admiral Canaris. In September 1940, the Three Power Pact was signed between Germany, Italy and Japan, which included an attempt at agreeing to intelligence co-operation. As the military objectives of both countries became more diffused, any attempt at co-operation in the field faded away. As Hitler attacked Russia in 1941 he tried hard to involve Japan in the conflict but she expressed little interest in joining him as her target was not Russia but the United States,

Britain and the Netherlands East Indies. Japan's negotiators did, however, express great interest in obtaining the keys to the American and British ciphers that Germany had broken and requested the codes from her allies. Flicke's account of the negotiation is that Japan was promised the codes that had been broken plus Germany's co-operation in other cryptographic matters. Berlin made the condition that Japan could have the cipher keys as long as she would enter the war against Britain and America immediately. Oshima sent several long telegrams to Tokyo, all of which were duly intercepted and decoded by the Germans so that Berlin knew exactly what their counterparts in Tokyo were thinking and saying. The Germans began to lose patience at the delay and gave the Japanese a 'now or never' ultimatum. The Japanese response was, 'not without the cipher, plus access to the telephone conversations between Churchill and Roosevelt and complete cryptographic co-operation'. Given those terms, the Japanese government agreed to declare war on Britain and the United States, and Japan began to plan the attack on the American fleet on 7 December 1941 in Pearl Harbor, just as Germany's offensive in Russia was grinding to a halt in the Russian snow.

A couple of months before the attack, the moderate Japanese Prime Minister Konoe resigned and was replaced by a more hawkish General Tojo, who requested that America and Britain withdraw the ban on the sale of oil and scrap iron to Japan. The embargo was effective and Japan became very short of oil, which America was sending to her ally China. Japan asked the United States if she could expand her trade into the East Indies, but both America and Britain saw this as a challenge to their authority in the area. The two governments found them unacceptable and a Japanese aircraft carrier task force and army began preparations to attack both America and Britain. Tojo did not know that at the same time America was beginning to read Japanese diplomatic despatches and guessing their intentions, although the coded exchanges were not specific about when and where the Japanese would strike. During the months leading up to the attack, American listening stations on Hawaii identified increasing radio traffic from Japanese naval ships as they moved south towards Hong Kong and Malaya. The intention to attack targets in South East Asia was fairly clear but the position of Japan's fleet of carriers was not. The Japanese fleet maintained radio silence for three weeks and Admiral Kimmel in command of the US Pacific Fleet was worried and wanted to know where they were. The first Americans to know that were a couple of soldiers trying out a new piece of ELINT kit called radar on a Sunday morning on the beach in the Hawaiian island of Oahu. (Radar was well known to the combatants in Europe, but it was just beginning to emerge as a technology in the American armed forces.) On their primitive screen they saw an armada of 350 fighter, bomber and torpedo carrying planes on their way to Pearl Harbor to attack and severely damage the American Pacific Fleet. The soldiers reported the sighting to the army's information centre and were told not to worry as they were probably a flight of navy planes about which the Centre had not been told. This was the first and most serious of America's inter-service lack of communications that would bedevil their intelligence performance in the Pacific.

Pearl Harbor was not the only attack made that day. The Imperial Japanese Navy went on to demonstrate their new skills in striking at unsuspecting targets, such as the Royal Navy. Territories in the East Indies, Malaya and the Philippines which were able to provide landing fields for their air fleet enabled them to raid Darwin and Broom in Northern Australia and sink a dozen ships. A British fleet of five ageing battleships and three aircraft carriers under Sir James Somerville were based in the Indian Ocean and received a report in April 1942 that Ceylon (modern-day Sri Lanka), was about to be attacked. Somerville's force was no match for the Japanese fleet of Admiral Nagumo and he knew it, so he kept his fleet well to the west of the island by day. Then as night drew in he steamed eastward to where he thought the Japanese would be hoping to take them by surprise. Somerville's information was wrong and, by 2 May, with the Japanese nowhere in sight, his ships were running out of fuel. He needed to return to base but then a report from one of his aircraft told the Admiral that the Japanese task force was sailing towards Ceylon but too distant for Somerville's force to intercept them. Two cruisers that he had put on detachment near the island were caught and sunk by Japanese aircraft and Colombo, the port and capital of Ceylon was attacked by over 100 aircraft, although the port was not put out of action. The British aircraft carrier HMS *Hermes* was in the port and when the alarm was raised she set sail without any aircraft on board but Japanese dive bombers from *Nagumo's* carriers soon found her and sunk her along with a destroyer and two tankers. The Japanese never returned to the Indian Ocean in any force after that but the damage that *Nagumo* did ensured that it would be some time before the Royal Navy was able to strike back.

Japanese signals intelligence did not go to war completely unprepared, only badly so. Over several years they had developed a coding machine that had progressed through a number of versions, beginning when the Japanese Foreign Office purchased an Enigma machine in 1936 for modification. This was the basis of the Purple machine. The ciphers of Purple run through the history of American cryptographics until the end of the war in the Pacific in a similar way to the Enigma machine in British cryptology. Indeed, the cracking of Purple was a joint Anglo-American affair, with the Japanese section of Bletchley Park making their contribution. This legendary Japanese coding machine was developed in successive stages, making the machine progressively more sophisticated with each version. This helped American decoders to understand the mechanism as they were able to study the progression of the stages as the red machine was cracked and then the purple one. It made the unravelling of the secrets of the Purple machine easier, although the Purple machine was not finally broken until 1943. Before that time, the Japanese order of battle and movement information was determined entirely from direction-finding and traffic analysis which was tracked in near real time. This was not as ineffectual as it might sound, as one major action achieved by Americans using tracking techniques located and identified a convoy of ships moving two Japanese infantry divisions from Shanghai to New Guinea. The convoy was intercepted by a 'Wolf Pack' of American submarines, causing the almost total destruction of a complete Japanese military formation and its sea transport.

The Pacific theatre was heavily naval in nature and the main battles were primarily between battleships and aircraft carriers slugging it out over vast stretches of water – but there was another aspect to it. American and British submarines led an assault on Japanese cargo ships sailing between the small islands to supply their garrisons, sinking a large number. They were slowly strangling the supply chain of the extended Japanese empire in the same way that German U-boats were trying to do for Britain. The Americans had a substantial submarine fleet at the beginning of the war consisting of three main types: the S Class (an ageing boat built in the 1920s and with little endurance for the vast distances of the Pacific); the huge cruiser submarines of the Nautilus type that were slow and lumbering in action and mainly used for clandestine work such as landing goods, supplies and men secretly on many tiny islands that dot the Pacific; and the Gato class. The Nautilus were the link to the Coast Watcher Service whose agents the US Navy placed in the thousands of tiny atolls across the ocean to report by shortwave radio the passing of Japanese ships. They were the Pacific's equivalent of Europe's radio agents and were the prey of Japanese troops seeking them out in their desert island hideaways. The Gato class of American submarine was the workhorse of this war of attrition, sinking hundreds of Japanese ships. Defective torpedoes made their task difficult, as it did with German U-boats in 1940, but the mechanism was rapidly improved. British submarines based at Trincomalee, where the author's father was based in the later part of the war, contributed to the onslaught. By the end of the war, the Japanese had lost over 8 million tons of shipping, of which 200 were warships, but the destruction of the merchant marine was one of the most important factors in the defeat of the Japanese empire.

In the Pacific the aircraft carrier was the trump card as it was able to take aircraft where land-based planes could never reach and changed the nature of naval warfare. Since Nelson's time and before, warships had to be in sight of each other to fight a sea battle. Carriers now enabled fleet actions to be fought in the Pacific largely by vessels that never even sighted each other. The first battle of the carriers occurred in the Coral Sea in April 1942 as the Japanese tried to cut the lines of communication between Australia and America by capturing Port Moresby in New Guinea. A seaborne landing force codenamed MO with a dozen troop ships and air cover from the light aircraft carrier *Shokaku* formed the task force. Two large carriers, the *Shokaku* and the *Zuikaku* lurked in the middle distance making up the Japanese fleet. The Americans had developed enough expertise in cryptography by now to intercept transmissions giving away Japanese Admiral Yamamoto's intentions to Admiral Nimitz commanding the American Pacific Fleet. A task force that included a couple of America's biggest carriers, the USS *Lexington* and the USS *Yorktown*, were ordered to intervene. Aircraft of both nations scoured the Pacific trying to find each other's fleet. The Japanese found an oil tanker with a destroyer escort, sinking both. American aircraft found the troop convoy bound for Port Moresby. The Japanese troop ships fled, abandoning their landing as the *Shoho* was sunk by American bombers along with the air cover on which the Japanese transports depended. The following morning, the two carrier fleets, which were 200 miles apart, clashed as they each launched their aircraft to attack each other. The American dive bombers hit the *Shokaku*, putting her out of action, while the USS *Yorktown*'s fighters were caught refuelling on deck as she was hit with a bomb. At the same time, the USS *Lexington* was attacked by Japanese torpedo bombers, causing large gas explosions which ultimately sank her. The Coral Sea battle was an opener for the big event which was to take place at Midway Island, so named because the island was midway between America and Japan. The action in the Coral Sea was technically a win for Japan but their losses deprived the Japanese of two carriers which could have proved decisive at the Battle of Midway, and prevented the Japanese invasion of Australia.

At Midway the Imperial Japanese Navy tried to ambush Nimitz's fleet, whose force had been weakened by the attack at Pearl Harbor, in a complex battle of movement in which carriers were the main participants. Yamamoto had eleven battleships and five aircraft carriers while Nimitz had no battleships and three carriers, one of which was the damaged USS *Yorktown*. The prize was the strategic island of Midway, but the Americans had one major advantage: their Radio Listening Service was able to determine the plans of Yamamoto and with this superior intelligence the Americans were able to strike at the heart of the Imperial Japanese Navy. With this insider knowledge, Nimitz was able to counter each of Yamamoto's moves, although sometimes by the skin of his teeth, by decoding the intercepts of Japanese plans. On 20 May, Yamamoto sent a long and complex directive to his fleet which the Americans were able to read. Code breakers were able to tell Nimitz that Yamamoto would mount a major attack on 3 June preceded by a feint attack on the Aleutians to draw the Americans away on the day before. The main attack would probably be Midway but this was uncertain because Yamamoto was using code names for place names. AF may have been the code for Midway, but Nimitz was not sure and he desperately needed to know. The intelligence officer of the fleet sent a signal in a code that he knew the Japanese had broken saying that Midway Island was short of water. A Japanese message was intercepted soon afterwards saying that AF was short of water, so the objective of Yamamoto's battle fleet was now clear.

The three American carriers, the USS *Hornet*, the USS *Enterprise* and the damaged USS *Yorktown* were at Pearl Harbor, and the engineers said that the damage to USS *Yorktown* would take three weeks to repair. Nimitz gave them three days. On the third day, the USS *Yorktown* was at sea with the other two carriers and Nimitz positioned them where Yamamoto would not have expected them to be, north of his main fleet. The Americans waited for Yamamoto to send his carrier planes to bomb Midway Island from his four carriers, *Akagi*, *Kaga*, *Hiryu* and *Soryu*, all of which had participated in the attack on Pearl Harbor. While they were away, the American dive bombers arrived over Yamamoto's carriers and sank the *Akagi* and the *Kaga*. Later the *Soryu* and *Hiryu* were both damaged. Yamamoto had lost all his carriers, giving him no alternative but to turn for home. The Battle of Midway was a complete victory for America and Japan had lost its aircraft carrier force, even though the USS *Yorktown* was torpedoed and sank after the battle. Midway signalled the high tide of the Japanese naval offensive in the Pacific but, following the loss of much of its carrier fleet, the Japanese were now on the defensive.

Following their great sea battles, the Americans went on the offensive on land as well as sea. A force of 20,000 marines landed on Guadalcanal where the coast watchers reported that the Japanese were building an airfield. Japanese bombers could threaten shipping and even the Australian mainland from there so a fierce and bloody battle was fought to contest the island. American marines badly dented the confidence and self-belief of the Japanese military and US forces began to take island after island in their progress across the Pacific. It was going to cost the US Navy dear; in a sea battle off the Solomons, the American fleet would lose a full squadron of its cruisers in actions in 'Iron Bottom Sound'. The Battle of the Coral Sea ended Japan's naval plans for a sea-borne landing in Port

Moresby so now the task was given to the Japanese soldier. They landed in New Guinea and advanced up the Kokoda Trail over the Owen Stanley mountains until they met the Australians, who proved that they could beat the Japanese in appalling jungle conditions as long as they had air superiority. As the Americans advanced slowly across the Pacific Islands, one clash in the Gilbert Islands was named Bloody Tarawa; 5,000 Japanese troops were entrenched in deep tunnels all over the island, severe losses were suffered on both sides and a savage battle was finally won by America's marines. There were only 100 Japanese prisoners left alive after the battle. Similar battles faced the British in Imphal and Kohima in Burma against the Japanese Fifteenth Army where they found conditions just as bad as the Kokoda Trail. A well-documented contribution to the great sea battles of the American and Japanese carriers was made by signals intelligence. Probably there was a similar contribution to the hard-won victories of the soldiers in their slit trenches in the islands but that does not seem to have been so faithfully recorded.

British troops had a long, hard fight on land against fanatical Japanese soldiers which brought the Japanese Army to a standstill in Burma. American marines took one island after another in the Pacific, finally meeting the desperate defence of the dedicated Japanese Army in Iwo Jima and Okinawa. At sea the Imperial Japanese Navy fought to defend the Mariana Islands unsuccessfully in the last great carrier battle of the Pacific War. Victory finally overcame what was the last desperate attempt of the Imperial Japanese Navy to stave off their inevitable defeat at Leyte Gulf, leading to the return of General MacArthur to the Philippines. Finally, sick of the blood-letting inflicted on his soldiers and sailors, President Truman ordered the dropping of the atomic bomb. It quickly brought a Japanese delegation onto the deck of the American battleship USS *Missouri* to sign a surrender document and end the global conflict on 2 September 1945.

Sicily and Italy

General Alexander had sent Churchill a triumphant message, 'The Allies had become the masters of the North African shore.' Now the best trained and battle hardened army the British had was eager for the fray; the question was, where should they attack and when? That was a question not only for the Allies but for Hitler as well; he had to man the defences of too many beaches that the Allies might storm as they whittled away his empire. As ever, the German intercept service was listening but the Allies had learned lessons in signals intelligence and kept a profound radio silence before landing in Sicily.

The German generals were learning to cosset their intercept companies because they were losing so many. As the Allies landed in Sicily, the German radio intelligence company based on the Island of Marsala was immediately transferred across the Straits to mainland Italy without loss of men or equipment. They resumed their radio watch from Reggio without a pause in their monitoring operation and, as the Allies advanced up the Italian Peninsula, they retreated back to Salerno and the Rocca di Papa regions. As the Americans and the British closed in on Field Marshal Kesselring's forces in Sicily, he evacuated his men, as the Führer had not issued an order of no retreat as he had done so disastrously in Africa. Over 100,000 of the field marshal's men with their full equipment crossed the Straits of Messina back on to Italy's mainland. The Allied advance into Italy caused Mussolini's government to fall; he was arrested and the ageing Marshal Pietro Badoglio became prime minister. The new Italian Govenment began to quietly discuss a separate peace with the Allies.

The quality of wireless transmission and evaluation of intelligence began to change about this time, as the German intercept services centralised their organisation. This radical change in procedure included directional finding techniques which suited the German operators fighting in Italy's mountainous terrain. Radio techniques shared between British and American operators had the effect that German operators could no longer differentiate between units, although occasional troop designations could give a unit's identity away. Allied radio communications were generally so good that, in early 1944, it became impossible to discern their intentions or their battle order. German operators took to studying RAF transmissions and codes which were always much more transparent than the army. Italian transmissions were also easy to read for the Germans so they were able to decode Italian Navy messages clearly indicating Italy's intention to make a separate peace with the Allies. A careless British transmission confirmed that terms had been agreed and that Italy was about to become a neutral nation. A month later Germany became an enemy as Italy declared war on her erstwhile ally. German troops immediately occupied Rome and, from then on, Italy was a German occupied country. The Allies' progress up the Italian peninsular was painful with many savage actions taking place on the way, one of the worst of which was the Battle of Monte Cassino. Long after the war the author visited Monte Cassino and, looking down from the terrace of the rebuilt monastery to the brow of a neighbouring hill, it was pure white with gravestones against the verdant green of the Italian landscape. The Germans sold Italy very dearly.

Counter Intelligence

As the Allies were fighting through the mountains of Italy and the Russians battling the Panzers in the Russian steppes, there was an altogether different enemy that the Germans had to face. The resistance fighters in the many countries that the Wehrmacht had invaded were dedicated to fighting the enemy in any way they could. They were driven by all sorts of motives, from the pure hatred of a German soldier or Gestapo agent, to the anti-fascist aims of the Communist insurgents. Some countries had a talent for international intelligence, its gathering and acceptance, while, as we have seen, others were less skilful, but almost every country worked well on counter-intelligence within their own borders. Enemy espionage and sabotage agents operating within the borders of Russia, England or even Germany faced effective home security organisations. It was a different story in occupied Poland and Czechoslovakia, then Norway, Denmark, the Netherlands, Belgium and Northern France in 1940; and, in 1941, the Balkans and a large swathe of Russia. Occupied countries sprouted underground resistance groups prolifically, always encouraged and supplied by Moscow Centre or London Control headquarters.

The most effective way to identify these resistance groups was to listen to their radio transmissions as they reported their findings or requested more supplies or people; the Abwehr and the Gestapo became very good at listening. Much of the following comes from the papers of Colonel de Barry who was director of German radio counter-intelligence from 1942 to 1945. The author met some of the colonel's lieutenants who had served with him in Berlin and who provided the papers quoted for a few cigarettes. Agent radio traffic was reported to one of five control centres:

- The Western net radio traffic control in London, overseeing British and a few American agents
- The Eastern net that controlled agents, partisans and spies from Moscow Centre
- The PS-net radio traffic control of Polish and Czech agents, some in London and some in Moscow
- The South East European net of British and Tito radio traffic in the Balkans, whose control was based in Cairo
- A few illegal transmitters based in Germany that were mainly reporting to Moscow Centre

The radio intercept branch of the German civilian police was responsible for locating unlicensed transmitters at the beginning of the war but was soon overwhelmed in 1940 by the volume of illegal transmissions that were being located. The Armed forces had to intervene and took over large aspects of the task but the co-operation, or lack of it between the Abwehr and the civilian police often led to jurisdictional conflicts and even the escape of enemy radio agents. The military method used to intercept long and medium wavelength transmissions was not very suitable for detecting agents' shortwave transmissions. By the end of the war, locating agents became even more difficult as the radios they used became increasingly sophisticated pieces of kit and were employed in increasingly cunning ways. Often three or four sets operated in a network taking turns in transmitting a message while each constantly changed its location. Agents chose their locations and transmitted with care when they could; generally hotels or large apartment complexes were used with several avenues of escape and several watchers on the lookout for danger. When a transmission location was identified by home security services, it would often be put under observation to identify further agents visiting the operator leading sometimes to the identification of agents visiting from other espionage networks. The arrest of wireless operators could be both difficult and sometimes dangerous; they were often located using short-range direction-finding sets that could be installed in motor vehicles with no antenna showing. To track an agent to a specific house or flat German security men used a miniature direction-finding set attached to their belts; such personnel were generally Abwehr members and not the home security forces so often police had to be summoned to make a legal arrest. This changed in 1943 when the Abwehr were given the right to detain suspects; the extreme caution of agents made immediate action necessary and counter-intelligence operatives needed powers of arrest. Such operations often turned into a gun fight, with armed radio operators being supported by his group who would probably have been keeping watch for him or her. There was a constant and serious loss of experienced counter-intelligence Abwehr specialists through death or injuries. The table below sets out enemy agents and their home control stations to which they were transmitting, as observed by Abwehr listening stations during the period 1943/44. They are broken down by country to show the intensity of agent activity in each one; the largest number being run by Moscow are located in Russia itself and must have included the partisans operating behind German lines.

Location	Russian Moscow	Polish London	Czech	British London	British Cairo
Russian Partisans	140				
Poland		20			
Czechoslovakia			4 ★ & 2 ★★		
Norway	2			15	
Denmark				4	
Holland	2			20	
Belgium	2			25	
Paris	3		2	30	
Western France				20	
Southern France	3	8		50	
Spain				10	
Switzerland	3	1			
Northern Italy				5	20
Southern Italy					8
Sardinia & Corsica				5	
North Africa				5	
Yugoslavia				5	25
Hungary	2	2		5	
Romania	5	2		5	
Bulgaria		3		6	
Greece					20
Istanbul	2	2			
Tiflis	2	2			
Cairo	1	1			
Total	**167**	**41**	**8**	**210**	**73**

★ London controlled
★★ Moscow controlled

Source – The papers of Oberst (Colonel) de Barry – Director of German radio counter-intelligence

In September 1944 Wilhelm Flicke was transferred from *Lauf* to *Funkawehr* (radio counter intelligence) in Zinna near Berlin to monitor agent and partisan signals traffic in the radio intelligence section of the Abwehr. At that time the service employed about 2,000 military personnel and some civilian operatives, in addition to which a similar number of 2,000 served in the radio intercept service of the civilian police. The Abwehr operated fixed radio stations in Germany, Italy, France and Austria with long-range direction-finding sets controlled by long-distance telephone circuits connected to OKW.

Arrests of radio operators increased as the war progressed, beginning with thirty radio agents arrested in 1941, and rising to ninety in 1942. As the number of radio agents operating increased, so did the arrests, which almost doubled to 160 in 1943. By 1944 the German Army was retreating and lost ground, so the number arrested dropped to 100 in 1944. From all these arrests, 410 cases were brought before tribunals and an additional 140 arrests were made based on information received from sources other than radio direction-finding. Each one of these numbers hides a story of terror, brutality, death and courage, although the Abwehr did not use force in the way that the SD did. They got very good results from informing prisoners how much they already knew and the suggestion that they would be handed over to the Gestapo if they did not talk would invariably be enough to elicit co-operation.

More than twice as many transmissions from agents' operating sets were intercepted as were actually located by direction-finding techniques in the period reviewed in the above table. These agents were never seized, of course. Also, probably twice as many illegal agents operated stations, although they were never identified by an overworked directional service. An unknown number of arrests were also made by the Gestapo in their counter-espionage activities as they never kept the Abwehr informed. An estimate made by the German counter-intelligence service of the number of agents operating radio sets of all kinds found in Europe, controlled by either London or Moscow, was in the region of 4,000. That gives a ratio of about one set operating for every member employed in counter-intelligence in the Abwehr, so their arrest rate was quite creditable. The German security services were overwhelmed by a rising tide of resistance organisations and their membership. German counter-intelligence operations were all on the alert and still looking for the traitor in the Führer's Headquarters feeding the Red Orchestra, but then they did not know about Bletchley Park.

Scandinavian Resistance

Denmark was invaded by the German Army in the first week in May 1940, using that country as a stepping stone to attack Norway. After some resistance, the tiny Danish Army surrendered in a couple of days. A German mobile intercept platoon was immediately installed in the Danish town of Aals to monitor Norwegian and British Army radio traffic coming from Norway, as already mentioned, but was more concerned with their transmissions to London. Strategically, Denmark was not very important to the Allies but they still needed to monitor Wehrmacht activities there so both English and Russian controlled agents began to be identified by the Abwehr in 1941. A raid took place a year later unmasking a Danish Comintern network in northern Denmark dedicated to the sabotage of railways; two other groups were observed but never identified, probably with a similar objective. Groups built up that were obviously intended as a support for the Allies as they advanced through Europe and the Danes began to reconnoitre military establishments and strengths. A programme of sabotage acts was being prepared but the speed of events in the spring of 1945 gave the networks little time to accomplish anything. It was not the same story in Norway.

After the invasion of Norway, the airways went quiet for a time, with Britain and France engaged in the Battle of France and the Russians not yet having entered the war. King Haakon and his government had wisely taken ship for England along with Prime Minister Nygaardsvold, who headed the government in exile in London. He broadcast regularly encouraging his people to passively resist the invading German forces. 'The armed conflict was at an end' he said 'but the spiritual conflict was just beginning', so the first agent transmitters were identified by the Abwehr very early in 1941. By the end of the year, radio reports were regularly being made to London from all parts of Norway and networks set up in a similar way to that used in France. Agent stations of Norwegian resistance groups were generally secure. They changed their locations frequently while only transmitting for a short time but the main safeguard was that German direction-finding was difficult mainly because of Norway's mountainous terrain. The Germans even used light Storch aircraft in direction-finding exercises but this only led to an occasional arrest and when one agent was caught another would invariably take their place. The ubiquitous agents provided a valuable service with a coast watch, one of which spotted the battleship *Tirpitz* at sea among many other sightings of vessels reported to London. The coast watchers organisation codenamed Scorpion was directed from Drontheim by Rolf Lystadt and shipyard director Bengt Groen who, with their network of agents, reported all ship movements along Norway's long coastline. In May 1944 an agent station on the island of Oncy was raided and a transmitter found in a cave that gave excellent views of passing ships but no operator was found. However, the documentation and equipment with which this look-out was equipped and others similar to it made it clear that it was virtually impossible for a German ship to enter Norwegian waters without a report going back to Britain.

Norwegian agents were also engaged in observing German railway traffic, but they became increasingly involved in sabotage of military and other installations.

Supplies of ordnance, equipment and agents came and went by sea as well as by parachute. A regular boat service known as the Shetland taxi service plied between Scotland's northern islands and isolated parts of Norway's coast 200 miles distant, becoming a part of the Norwegian resistance movement's saga. Similarly, the audacious attacks on the Nazi production of heavy water for atomic bomb research, perpetrated by resistance men and women and by RAF raids, has passed into legend.

In addition to British sponsored resistance in the country, there was also a strong Russian intelligence initiative mainly centred on Northern Norway and around the port of Narvik. Their agents, controlled by Murmansk station in Russia, reported German shipping movements so well that barely a ship could put to sea without a Russian aircraft from Soviet territory appearing to attack it. The sets they operated used wavelengths of 80 to 100 MHz, at a low output, and the transmission was difficult for German monitors to pick up. Nevertheless, the Abwehr caught one Russian agent with his radio intact and proceeded to play a radio game with Murmansk, hoping to guide other agents being put ashore by submarine carrying their shortwave receivers. Using the captured set, the Germans lured one Soviet submarine into a coastal area and sunk it while capturing the Russian agents as they landed. This was one of several severe blows to the Russian network in Northern Norway from which they found it difficult to recover.

France

Before war was declared in the summer of 1939, British intelligence had begun to set up a network of radio agents in Western Europe, but within a year their radio stations were overrun by the German Army. Established sleeper agents were left to act as the basis of an agent network that began its expansion in the following year, with newcomers arriving by night in light Lysander aircraft or by parachute carrying the newly developed shortwave transmitters with which to transmit to their home controller. The German radio defence organisation started monitoring agent radio traffic in France late in 1940 and also used its directional location service to identify a British control station. Wilhelm Flicke tells us that direction finders located transmissions north-east of London, manned mainly by French, Dutch and Belgian operators. Could he have been talking about Bletchley Park, based outside what is now Milton Keynes? Transmissions to agents from this location and responses were not very secure at this time; they probably felt they would not be located by directional tracking but they were wrong. Radio agents observed in both occupied and unoccupied France numbered about twenty-five in July 1941, growing to fifty-six stations by October with networks beginning to expand all across Europe.

Meanwhile, the British Secret Intelligence Service had nurtured the beginnings of a network in France already described. The intention was to define the order of battle of German forces throughout Europe. To achieve this, their methods were those of the traditional spy ring. Their agents had to observe and record enemy establishments, movements and strengths and then report them back to their control in London without the enemy being disrupted or maybe even knowing that they were being watched.

In July 1940, the Special Operations Executive, or SOE, was formed at the urging of Winston Churchill with the very different brief of 'Setting Europe Ablaze'. They constantly stirred up German security services by sabotaging rail and telephone communications in a way that kept them jumpy, active and alert. As can be imagined, with such different agendas the two organisations did not get on well; SIS looked down their noses at SOE as amateurs and SOE retaliated saying that the SIS were 'sticks in the mud' and worse.

All this time, governments in exile were trying to do their own thing with their agents in their own country while the British government wanted to maintain at least a modicum of control over the sometimes diametrically opposed intentions of their guests. The Free French government in exile created their own intelligence service and also made up their own rules under the leadership of General Charles de Gaulle, a prickly individual who did not get on with anybody. De Gaulle's secret service organisation, headed by Major André Dewavrin, was initially known as the Service de Renseignements (SR). In April 1941 it was changed to the Bureau Central de Renseignements et d'Action Militaire (BCRAM) and then again in January 1942 it was shortened to the Bureau Central de Renseignements (BCRA), by which name it became best known.

With many operatives in almost every Western European country having their own agenda and objectives, results were predictably chaotic. Even worse were the resistance groups operating in the field as they were often split within themselves, divergent and splinter groups from Communists to ardent nationalists who were at war with each other almost as forcefully as they were with the enemy. The intelligence services operating in Europe and beyond were not a homogeneous bunch, particularly when pitched against the low cunning of the German Abwehr and the brutality of the SS, so they made some blunders. Fortunately for the British and their allies, the Abwehr was also working in a positively dysfunctional intelligence community with the Gestapo vying to take over not only counter-intelligence but also, as time went on, acting as spymaster on the international scene as well. Nevertheless, the German record of counter-intelligence was a good one, particularly against the French and Dutch underground movements.

De Gaulle's intelligence war was a secret from the British but not the Germans and his BCRA was organising a secret army known as the Organisation de L'Armée Secrète (OAS). SOE were supplying arms and support to the French, of which the OAS got its share. General Charles Delestraint, whose code name was General Vidal, was in command and he built up a structured organisation that stretched into the furthest regions of France. Vidal established a command structure that organised service instructors in training programmes for different aspects of the Armée and set up arms dumps and wireless centres to supply it. It ran a recruiting campaign to expand its membership and de Gaulle was proud of his achievement. Unfortunately, they recruited some members of the Abwehr, some of whom were even promoted to staff level of the organisation. The Abwehr knew more about the Armée than General de Gaulle did, so when the Germans were ready, they arrested the lot. General Vidal was shot and regional leaders were arrested by the hundreds and many members driven to live underground. It had an effect just as profound as the better known takeover of the Dutch resistance movement in the North Pole operation. The difference was the Dutch had a public investigation after the war of their North Pole intelligence disaster, the record of which they printed in several leather bound volumes. The French made no such effort with the OAS – maybe there was much to hide.

For the purpose of strategic resistance in 1943, the failure of the Armée Secrète had far reaching effects on future Allied plans for the invasion of France and acted as a considerable lift for the morale of the Abwehr. In London De Gaulle had been treated to a blast of the obvious – clandestine espionage organisations needed to be small and discreet as the German counter-espionage organisation was a formidable enemy. As a result of this disaster, the central emphasis of activities of Allied espionage networks shifted to unoccupied France with Spain as the backdoor route into the country. Consequently control stations for radio communications were set up in Madrid, Barcelona and Cartagena for agents in both occupied and unoccupied France. As a result, the Abwehr began its Operation Donar in co-operation with the French Vichy police, raiding radio emission locations in unoccupied French territory. According to Wilhelm Flicke, the so-called Vichy Wave, an influx of agents in unoccupied France, caused the Germans much trouble as various transmissions were observed on wavelengths often used by the Vichy government. After D-Day and, particularly, after Operation Anvil (the landing of Allied forces on the south coast of France), co-operation with the police and their informers dwindled to almost nothing. In spite of this, Abwehr counter-intelligence activity continued and the remains of the espionage ring that had been led by Vidal was broken down even further, dealing another blow to the movement in that area. During this time, German counter-intelligence had what Flicke called an unusual experience. Several networks had been penetrated, including one called Alliance in Marseilles which contained a young woman of unusual energy, intelligence and beauty. She was picked up in a routine raid but could not be identified by the local security men. She was put in the charge of the local Kriegsmarine unit to guard her. Her captor took an immediate liking to her and spent a pleasant night with her but, being a cautious fellow, he locked up all her clothes to guard against unpleasant surprises. Meanwhile, Abwehr officers had seen photographs of the lady and realised that they had caught a major agent but when they went to the naval unit to collect her next morning, they found that her captor was sound asleep and that the lady had disappeared into the night stark naked. She was never heard of again – at least by the Abwehr.

A close connection between agent activity in France and preparations for landings became noticeable. Just before the Dieppe Raid, the radio traffic increased as did the number of agents identified. The radio traffic also increased markedly all over France as D-Day approached, with the populace willing to be more robust in their attitudes to the German invader. In some areas a single vehicle of the German radio defence organisation could not safely travel the roads. This was particularly so at night when direction-finding vehicles would be on the prowl trying to use their directional equipment to identify radio transmissions. Younger people began to disappear in 1943/44 into the camps of the Maquis, the French Resistance fighters in Southern France. These groups included numbers of fugitives from the Gestapo as well as patriots who wanted to fight for their country. Supplied mostly by air with light arms and communication equipment, these groups used hit and run tactics and were a great thorn in the side of the occupying forces.

American agents began to be employed in France from 1942 onwards and, using much the same radio security methods as the British, they were located largely in the south of France before Operation Anvil, as well as Spain. They were controlled and supplied mainly from Algiers in North Africa by the American Office of Strategic Services (OSS), brilliantly led by Major General William J. Donavan, or 'Wild Bill' as he was known to all. He modelled his service on SOE and played a large part in the invasion of France, particularly on its southern coast, and also Sardinia and Corsica. After D-Day the Abwehr left France with more of a whimper than a bang as General Leclerc and his Free French soldiers raced through the tree-lined roads in the high summer of 1944 to the suburbs of Paris. Then it began.

The crowds came out in their hundreds of thousands to welcome the Free French heroes who had not seen home for five years. Prominent among the crowds were Frenchmen proudly bearing their Free French armbands of the resistance, some of whom had never been in the resistance. The enduring image on the newsreels that summer was the dense swirling crowds progressing slowly up the Champs-Élysées towards the Arc de Triomphe. Leading that exuberant crowd was the tall, gaunt, stiffly proud figure of General Charles de Gaulle. The mood of wild and abandoned joyfulness turned to panic in a moment as a German sniper fired at the crowd. Shots rang out and every person in that great crowd ran screaming for cover all with the exception of one, De Gaulle, who marched unhesitatingly on towards the Etoile and the Arc de Triomphe as all about him crouched behind cover. This was one of the great dramatic moments of the war.

Belgium

No agents' transmissions had been intercepted in 1940 by German direction finders in Belgium, even though there was evidence that some were active. Early in 1941, the arrest of an agent named Martiny with a transmitter he had hidden under coal in his cellar changed all that. There was enough documentary material to give a good picture of his espionage activities and an agents' radio net that spread into France, Holland and Luxembourg. It transpired that there were a number of *kappelen* or agent groups in Belgium as the country was being used more and more as a base for operations by the British. A number of arrests confirmed the cross-border links and the first indication emerged about a Belgian secret army after some careful direction-finding. This organisation was a large network mainly of former Belgian Army officers that SOE had recruited to use in its operations. The Abwehr swept up much of it, proving once again that networks were safest when made up of small, discrete cells. Resistance groups worked quite well on the whole in Belgium and acted as a hub for groups in Luxembourg and also had connections in the Saar region but it was all to be affected by the most extraordinary drama about to be played out next door in Holland. The activity of agents in Belgium, as in France, reached its peak just before the invasion as all movements of German troops and supplies, in fact the details of every Wehrmacht headquarters down to company level in Belgium, were reported to London Control. All held their breath as the tanks and men of the Allied armies raced across the border with France, reaching Brussels to be met by more joyful crowds. The people of of this historic European capital were so pleased to see their liberators that they called a whole section of the city after General Montgomery, who rolled into the city with a cavalcade of flower-bedecked tanks. The British had seen so many joyful liberations on the newsreels that they were getting quite blasé about it and looked forward to similar scenes in Holland – but it was not to be.

Holland and the North Pole

In German-occupied Holland in 1941 an espionage network was being built up by the exiled Dutch government's secret service, or *Inlichtingendienst*, in London. Two agents were parachuted into Holland with a transmitter each; the German intercept service immediately spotted their transmissions reporting their safe arrival back to London Control. Major Hermann Giskes was the newly appointed Head of Abwehr counter-intelligence in the country who knew little of the Dutch espionage service at that stage. When a Dutch collaborator gave him a report saying two SOE agents had recently parachuted into Holland to set up an agent network he said, '*Gehen sie zum Nordpol mit Ihren Geschichten*' (go to the North Pole with your tales). The German counter-intelligence Operation North Pole started in spite of his disbelief and the first transmitter was located in The Hague by a direction-finding unit. The men of the unit were strictly forbidden to make an arrest themselves but waited for the counter-intelligence people to make the arrest and cover themselves in glory. Consequently, the direction unit stood by as the radio operator and his two assistants escaped over the roof of the row of houses in which they had been hiding. The second transmitter was identified a couple of days later by members of radio defence, who immediately arrested the operator with a suitcase of over 800 intelligence reports ready to transmit to England. The Abwehr now became aware of how active the espionage network was and how it reached Holland and into neighbouring countries. A successful co-operation began between the Abwehr and the security arm of the SS. This was unique as the organisations were normally at daggers drawn. Ignoring the feud between their services, Giskes and Joseg Schreieder of the Gestapo's security police arrested a dozen more agents and their radio sets. This led to the downfall of a major network of agents in Holland and laid the foundation for Germany's most successful counter-intelligence operation of the war, which Giskes recounts in his book *London Calling*.

Eleven of the captured Dutch transmitters were taken over by Abwehr supervisors, forcing operators to co-operate by transmitting to London without raising the suspicions of their controllers. If they did not co-operate, the agents would be handed on to the Gestapo. Agents were trained in the event of capture to incorporate or omit specific items in the text of a message to act as an alarm signal to their controller that they were transmitting under duress. Their German captors were unaware of these prearranged danger signals so captured operators sent their messages with the deliberate mistakes, such as an error in every sixteenth letter of the transmission. That particular measure was later discarded as bad reception could make accidental mistakes in the code and confuse the control operator but a range of others symbols could be incorporated into the text to sound the alarm. Amazingly, these covert signals were transmitted in regular messages to London but were completely discounted and ignored by control supervisors. Networks in Holland and related ones in France and Belgium were betrayed as a result and many agents with thousands of tons of supplies meant for the resistance were dropped straight into enemy hands. The performance of sabotage invented by Major Giskes for his captured band of radio agents, including some real events such as the sinking of an old ship for which the Germans no longer had any use, delighted London Control, who thought it offered 'proof' that the agents were doing their job. Giskes even arranged that details of the 'atrocity' be published in the German press that he knew British intelligence would read. The operation, known to the Abwehr as *Englandspiel*, or England Game, was a triumph for the Abwehr and cost not only the British intelligence networks the lives of dozens of agents but also some aircraft and their crews as they delivered tons of supplies into German hands. More than that, it narrowed the options for the Allied invasion of the continent as General Sir Alan Brooke, military advisor to Churchill, had favoured Holland as the site for D-Day landings. It was obvious that they could not have landings in a country that did not have an active resistance network to prepare the way for them.

The game could not last forever; it was a Dutch agent, Pieter Dourlein, that put an end to it. He fell into the hands of Major Giskes, the Abwehr's greatest spy catcher, a few minutes after landing in a tree near Ermelo in Holland. He found himself treated very well, along with the other fifty or so agents who had also parachuted into captivity. He was not handed over to the Gestapo to be tortured and brutalised as he expected, but housed comfortably in a converted seminary. He was interrogated intensively for any information to help feed and maintain the deception of a negligent London Control which had cost so many lives. Dourlein did not stay long in captivity but made his escape from the seminary to travel across France posing as a Dutch worker helping to construct new airfields for the Luftwaffe. He made a formidable journey through Belgium and France to Switzerland, then to Spain and finally ending up in Gibraltar to tell the British secret service his story. Giskes had got there before him and transmitted radio reports to London Control of Dourlein's co-operation with the Nazis. When the

Dutchman got to London they did not believe him and promptly locked him up in Brixton Prison. Someone was listening, though. His story reached the head of MI6, Stewart Menzies, who was told that his service had an efficient network of agents in Holland. Menzies, however, already had his doubts about the security of the network. On checking Dourlein's story, the existence of the whole network turned out to be illusory. Major Hermann Giskes had been responsible for the capture of a great many more 'turned' agents than the British XX Committee who had captured and 'turned' all the German agents entering Britain.

The *Englandspiel* was a very fortunate operation for Canaris as the Abwehr had incurred Hitler's rising displeasure far too often and this bright spot in his performance meant that the Admiral would still be secure in his job, at least for the time being. The game in Holland had been played faultlessly by Giskes for nearly two years but was now coming to an end so he decided to go out on a high. He sent a message to Majors Blunt (probably already spying for the Russians with Burgess and Maclean) and Bingham, the two officers in charge of the Dutch SOE Station responsible for building and supplying the agent's network.

Gentlemen,
Recently you have been trying to do business with the Netherlands without our assistance STOP We think this rather unfair in view of our long and successful co-operation as your sole agents STOP But never mind, whenever you come to pay a visit to the Continent you may be assured that you will be received with the same care and result as all those you sent us before STOP So long STOP

Thus *Englandspiel* had come to an end and the fifty agents Giskes was holding in the seminary were sent to a concentration camp where only two survived. Inquests into the failure of SOE have gone on until this day; the most thorough one was carried out by the Dutch government. Its findings were published in six beautifully bound volumes with the unsurprising (and probably true) conclusion that no treachery on the part of the Dutch participants had contributed to the disaster. The main question was never answered. How could the *Englandspiel* have gone on for twenty months? This critical question was asked by the Dutch Commission of Inquiry who came to London to find out why the security checks in the coded messages had not rung the alarm bells that they should have done. They asked for copies of the thousands of messages received by SOE from Dutch, French and Belgian agents over those fateful years. The messages, they were told, had been destroyed shortly after the end of the war. Even Giskes, when giving evidence in the inquiry, said he was perplexed by the disastrous lack of control. A standard practice that should have been carried out was to drop an agent into Holland without announcing his arrival to make sure the operation was going well but this was never done. The only bright spot for SOE in London was that the *Englandspiel* operation did not reveal the plans for D-Day. For Major Giskes and the Abwehr it was a triumph of their counter-espionage service. Giskes and Dourlein both survived the war and wrote books about their extraordinary experiences, and Giskes was last heard of living in Bavaria having joined the Gehlen intelligence organisation.

The British radio network in Holland took some time to recover after the *Englandspiel* disaster but a working network of agents reporting on German military formations with their strengths and movements was essential to the planning and direction of the forthcoming invasion. A new organisation, which the Abwehr called *Ordre-Dienst*, was identified in the summer of 1943 and was being slowly and painfully built up into a new net of agents. Three transmitters were identified in Amsterdam and Rotterdam and the principal agent in Amsterdam, von Borsum-Buisman, was soon arrested. From his papers German intelligence found why *Englandspiel* was blown and how Dourlein had reached Gibraltar. Further arrests were made in Zaandam and fourteen radio sets and some papers were seized but the *Ordre-Dienst* group was not broken up, indeed it flourished. It was not entirely untouched, though, as a station in Leiden reporting on V1 flying bomb installations and another in Rotterdam with a complete plant for producing forged papers were all discovered. By the time D-Day came there were three groups operating, *Raad van Verzet*, *Knock-Ploegs* and *Ordre-Dienst*, which together formed the Delta triangle network using about twenty radio stations to cover Holland. In September 1944 resistance personnel were formally recognised as members of the Dutch armed forces by Prince Bernhardt of the Netherlands and afterwards they were used by General Eisenhower in his offensive against the retreating German Army. They were ordered to protect water and electric power plants and the dock installations in Rotterdam, Amsterdam and IJmuiden. Agents were also instructed to seize certain bridges and prevent the Germans blowing them up, and mark minefields for advancing Allied units, identifying themselves by prearranged passwords. Agents also worked to assist troops and Dutch government officials to meet and talk with harbour administration, canal and river regulation managers. They were asked to meet at Eindhoven after the liberation to arrange the future governance of Holland. They came in their hundreds, some of them even by canoe, to regain control of the infrastructure of their country as it was freed by the Allies. It is clear that the Delta Network did much to repair the damage that *Englandspiel* had done and in doing so had helped the Allied cause and the resurgence of the Dutch nation considerably.

11

The Second World War – The End

D-Day

In the spring of 1944 German intelligence identified the move of elite British and American divisions from the Mediterranean to England. The American 82nd Airborne Division which had been serving in Southern Italy had not been heard of by the Abwehr for about three weeks. A transmission from a station located in England referred to a soldier whom a girl in America was pursuing as father of her child. The soldier's number quoted tallied with the code designation of the 82nd so Abwehr communications reported the suspicion that the division was now in England. This information was treated with disdain by the High Command as no ship transports capable of moving a division had been noted passing Gibraltar. It was mockingly suggested that they had been transported by submarine but a radio watch was maintained and evidence was soon found to confirm that the division was now in England. The Germans were able to keep track of Allied vessels passing through the Straits of Gibraltar with a sensor that they had installed on Spanish territory. However, what the Abwehr did not know was that the Allies had the technology to negate the German sensors, so that British ships could pass through the area invisibly. The British only occasionally used this technology for the passing of important shipping movements; in particular, they used it to mask the passing of the great Torch armada and for large D-Day preparations, of which the shipping of the 82nd Airborne was a part. The 82nd Airborne was one of the first units to parachute into France on D-Day, with its story told graphically in the TV drama *Band of Brothers*. German sensing equipment followed the landings in Normandy from D+2, so the High Command was able to identify 95 per cent of the units embarking in England. They were able to calculate accurately the strength of Allied forces in the bridgehead up to D+7. This information was derived entirely from radio intelligence as aerial photography was not available to the Germans due to the Allies' total domination of air space.

Wilhelm Flicke gave an account of intercepted radio activity during many landing exercises, giving an indication of projected landing procedures, although they could get no clue as to their time and place. No change in radio patterns was observed until the day before D-Day and no radio deceptions were observed but, according to later reports, that could have been because the first wave sailed at very short notice. The truth was that on the day of the assault a radio plan for landings on D-Day on 6 June 1944 took the German High Command by surprise due to absolute radio silence. German radio stations on the French coast had planned their procedure to meet the effects of the expected landings and was able to handle retirement from the coast and additional interception workload well. All monitoring of less important areas such as Ireland, Portugal, Spain and Brazil was discontinued by all units and the transfer of traffic intercepts proceeded smoothly. Enemy air attacks damaged communications and reduced the speed of intelligence reporting but an already rehearsed plan enabled transmissions to continue to be received at OKW. Signals traffic was re-routed as one intercept station after another was knocked out by enemy action. The experiences of the intercept intelligence defence in the papers of the German General Praun commanding the listening service in Normandy are worth recounting:

Immediately after the landings long range intelligence at first produced only minor results. The Allies did not wish to give any clues to our radio intelligence and therefore restricted radio communication, moreover, the short distances within the beachhead areas probably permitted the issuance of verbal orders and reports. In addition, the enemy was able to use telephone connections, which were not disrupted by any Luftwaffe interference (because there was none). The expansion of the beachheads resulted in the transmission of so many radio messages that a fairly clear picture of the enemy situation was soon obtained and an even greater wealth of information was provided by short-range radio intelligence and divisional combat intelligence. The signal officer for O B West moved his short range intelligence company nearer to Caen to improve his intelligence operations. Communications intelligence about forty-eight hours after the first landings was able to submit a list of most of the enemy divisions including information on the enemy army group command structure. The post-war press gave much attention to the opinion expressed by General Jodl, Chief of the Armed Forces Operation Staff, who said that a second landing was expected north of the Seine and therefore the German reserves and the 15th army stationed in that area were not immediately committed to a counter attack.

Communication intelligence did not support this assumption and Chief of the Control Centre of Communication Intelligence West, was asked to express his personal opinion on this matter during a conference of the Western intelligence branch. He said that a comparison of the number of units already recognised with those previously identified in Great Britain permitted the conclusion that most of the allied forces had already been landed and that the remaining ones were insufficient for a second landing. Any uncommitted units would be needed to feed the current battle. This opinion was shared by the Western intelligence branch, but was in contradiction to that of the armed forces operation staff. The estimate of the situation was given some validity by the fact that a short time after the beginning of the invasion a British landing craft had been captured near Boulogne. However, it became obvious that this enemy craft had lost its way.

When, during the next few days after the beginning of the invasion, the Allies created the impression of a second airborne landing by dropping dummy paratroops over Brittany at night, communication intelligence offered evidence to the contrary because of the complete absence of enemy radio traffic in the alleged landing area.

Throughout the war, General Jodl as well as Hitler himself frequently showed a lack of confidence in communications intelligence, especially if the reports were unfavourable. However, orders were issued as early as the time of the Salerno landing that all favourable reports should be given top priority and despatched immediately regardless of the time of day. Communication Intelligence West was required to furnish a compilation of all reports unfavourable to the enemy derived from calls for help, casualty lists and the like. When, during the first days of the invasion, American units in particular sent out messages containing high casualty figures, the OKW was duly impressed. In contrast, the estimate of the situation prepared by the Western intelligence branch was absolutely realistic and in no way coloured by optimistic hopes.

As the front in the west stabilised after the break-out from Normandy, Wilhelm Flicke's unit began an intensive analysis of the tactical use of radio telegraphy to learn the groupings and intentions of the Allied armies as they advanced. The Canadian First Army was extremely cautious in their use of radio so the Germans could learn little from them. The British Second Army was a little more lax in its communications but the resultant intelligence was very meagre. The Free French First Army was very circumspect at first but, after the liberation of Paris and as it advanced through France, it became less reticent and gave Flicke and his colleagues some increasingly good interception results. The American First, Seventh and Ninth Armies were unusually quiet, particularly the Ninth, and gave the listening posts so little in signals traffic that on many days German intercept had little to report. The American Third Army, who were less well trained and committed an exceptional number of radio communications security errors, was more forthcoming and gave Flicke's listening posts much useful information.

The Battle for France and the Bulge

The Allied breakthrough at Avranches forced German communications intelligence units to keep moving their mobile units back as the German Army retreated, so disruptions to the interception service were frequent. That was also true of the German Army group facing US General Patch as his French and American divisions landed on the beautiful Cote d'Azur during Operation Anvil. In three days Patch's men had created a bridge-head 40 miles wide and were driving hard up the Rhone Valley pursuing the retiring German formation that had been facing them. They were withdrawing fast at Hitler's order before the Allies pressing forward from the Normandy beaches could cut off their retreat and the intercept units had a hard task keeping ahead of Patch's Americans. The objective of the landings had been to relieve the pressure on the Normandy beachheads but in reality it did the reverse as German troops retired from the south of France. Operation Anvil failed in its objectives because it gave the German commander in Normandy additional troops and reinforced the formations facing the Allies, which made the situation somewhat more difficult for them.

After the Allied break-out in Normandy, trained Abwehr intercept personnel were increasingly drained away to serve as frontline troops. In addition, the disruption of constant movement made things difficult for signals intelligence. The re-siting and calibration of equipment as the army retreated provided a stern challenge for the German listening service, but it continued to provide reliable intelligence to its commanders. German intercepts were able to predict any attack of divisional strength or more up to five days in advance as American field ciphers were becoming compromised and their coded messages were decrypted in a few hours. The British Army cryptographic service was still secure, although RAF transmissions continued to be careless. The best communications intelligence obtained was derived from routine interception of air liaison officers' reports on the state of readiness of aircraft and crews. This gave indications of Allied intentions as aircraft were prepared to give cover operations and support a potential assault, which proved very useful. The morale of radio operators on both sides had begun to suffer for some time due to increasing artillery fire creating what was termed radio psychosis. It was first observed among German operators during the first days of the landings when enemy fire from both land and sea was very intense. Operators believed that as soon as they began to transmit they were being located and homed in on by enemy artillery. It took some time for officers to convince the men that shelling was as intense all along the line and they had not been singled out. This may have been at least a part of the reason that radio silence was maintained by the German Army as they prepared the Ardennes Offensive called *Wacht am Rhine*, or Watch on the Rhine.

Indeed, radio silence was so well maintained that the offensive caught the US Army by surprise. Some American units were found sleeping soundly but that was not the only reason for the German success. German radio intelligence had scored probably its last great success in the war as the assault started. An American military police network was picked up by the listening post in the Ardennes so they were able to listen to the radio transmitters of MPs (Military Police) positioned at every major road intersection along the front. This enabled operators to assess the exact strength and movement of British and American units as they countered the German offensive. Army formations in the battle did not have the fire power to take advantage of the intelligence, however, but German intelligence services still operated comparatively well in covering events up to the end of the war. The American advance to the Remagen Bridge and the story of the crossing is well documented and so was General Patton's armoured division's objectives in Germany as they crossed the bridge into Germany. Field Marshal Kesselring estimated at this stage of the war that almost all of his intelligence regarding an enemy order of battle and objectives came from radio intelligence as there were so few prisoners to question and no aerial photography available at all due to total Allied air superiority. The overwhelming strength of the Allies in all military resources and the lack of them in the German Army meant that German communications intelligence became of less and less use to commanders at any level because they could not act on it and so it became completely theoretical.

Operation *Ostereiaktion* or Easter Egg was a project hatched by the OKW to the amusement of Admiral Canaris and his Abwehr colleagues. It consisted of caches of arms and explosives buried for the use of saboteurs behind the lines of the enemy to hinder their advance. The boxes were mainly assorted items of ordnance dropped by British aircraft for the resistance and captured by the Abwehr. The boxes were buried at a certain depth near a specific landmark for future recovery mainly in sites in northern France and southern Germany as well as Poland and Czechoslovakia. The speed of advance of both the Russian and the Anglo-American forces on both fronts caused the operations not to be implemented. The author assisted in recovering some of those containers whose explosive content had become unstable after being interred for so long.

Middle Europe – Poland and Czechoslovakia

On the Eastern Front in 1944 the German Army was being destroyed by the Red Army as they swept all before them. The German Army Group Centre collapsed and organised partisan groups, directed by Moscow Central on their shortwave radios. Mayhem was created in the rear areas with over 40,000 separate acts of sabotage, destroying rail and road networks, communications and supply dumps. The routes of withdrawal for German troops were blocked, which contributed effectively to the disaster that overcame that large army formation. Partisans targeted mobile listening units thus depriving German commanders of much needed intelligence. Wilhelm Flicke was transferred from Lauf Listening Station, no longer needed to monitor the near east, to the relative safety of Zinna to survey partisan activities in Poland to Yugoslavia. As the Red Army crossed the Polish border, the Poles took up the fight as partisans to devastate the logistical infrastructure of supply behind German lines as the Russians had done. German units rated the Polish partisans facing them very highly as they were well equipped and supplied with arms by the OSS in Cairo through governments in exile in London.

Meanwhile, SIS and SOE still showed a determinedly spiteful lack of co-operation because the difference in their objectives set them so far apart. Now SOE was quarrelling with itself and splitting into two different organisations, the home section based in Baker Street in London and the other Cairo. The way in which the two parts worked set them apart, with each SOE office making totally different demands on its agents and organisation in their own region. The London office in Baker Street maintained the original objective that Churchill bequeathed it of stirring up trouble, mainly in Western European countries for the 'Occupying Forces', as the Germans called themselves. They were not bothered about gathering intelligence about how many soldiers the Germans had there; that was for that toffee-nosed lot in the SIS who had never blown up a bridge in their lives. On Cairo's patch in the Balkans, SOE agents were generally a part of a well-armed partisan group whose need to blow up bridges required some intelligence work to do it. Such matters as the lay of the land, the strength of the nearest German garrisons and others that might be of importance to them needed to be reconnoitred before a group of desperados attacked a well-armed emplacement in the approved partisan style. This all became obvious even to German intelligence who had fooled the London SOE office into making one blunder after another in Holland, France and other areas of operation while the Cairo office, which ran over half the agents in the field, were in German intelligence estimation, by far the most efficient.

Comparison between the two offices was most clearly illustrated by their attitudes to the resistance movement, or rather movements, in Yugoslavia. Colonel Milhailovic led a largely Serbian force of ex-army soldiers, while a large and growing Communist group under Josip Broz, later to be known as Tito, led another. Not a lot was known about either of the groups and their effectiveness as fighting units except that they hated each other just as much as they did the Germans. Cairo took the view that the group that engaged the invaders of their country in the most deadly way was the one to support while London, which was influenced by King Peter of Yugoslavia and his government in exile, favoured Milhailovic because he was not a Communist. SOE in London sent one of their senior members to Cairo to impose their view, but it soon became obvious to everyone that Tito was more effective at attacking the Germans. Churchill decided to send Brigadier Fitzroy MacLean out to report on Tito, who MacLean decided had earned the support of the British. Huge air drops of arms and supplies were made to Tito while Milhailovic, who was found to be co-operating with the enemy, got none at all. All this was monitored by Wilhelm Flicke's listening stations while directing German units in the increasingly bitter fighting against Communist partisans. They had disarmed the Italian occupying forces and Tito was using their weapons more and more effectively. All this fed Hitler's fear that a landing was about to be made by the British in either Greece or on the Adriatic coast and so held down a number of German divisions that could have been used in Italy or France. Tito's Communists were such an effective fighting force that they caused the occupying German invaders to retreat from Yugoslavia without the aid of an Anglo-American or Red Army. The British paid the price for supporting the Communists soon after the war when Churchill ordered the British Army to fight Communist forces trying to take over the Greek government.

The Bomb Plot

By 1943 it became increasingly apparent that Germany was losing the war. Hitler and his Nazi compatriots were unable to admit it to themselves or others, although there were an increasing number among the Wehrmacht that did. A number of senior army officers had been talking about assassinating the Führer even at the high tide of the German Army's advances into Europe and North

Africa in 1941. Indeed a number of ineffective attempts had been planned and even attempted but failed because of a technical difficulty. One such was a bomb planted in Hitler's plane that failed to go off as he flew over Russia. One centre of disaffection was the Abwehr where Admiral Canaris was involved in discreet talks about an assassination while his Chief of Staff General Hans Oster was up to his neck in planning an attempt. As the world knows, on 20 July 1944 a bomb was put under a desk where Hitler and his generals were still planning the domination of Europe, even though the roof was falling in on their world.

Colonel von Stauffenberg primed the bomb in his despatch case and left it under a table in Hitler's HQ at the Wolf's Lair in Rastenburg in East Prussia where the Führer was holding his planning meeting. Ten minutes later acid ate into the copper detonator and a large lump of plastic explosive blew the walls of the wooden building out. Von Stauffenberg saw the explosion and was convinced that the Führer could not have survived the blast and immediately flew to Berlin to meet the other conspirators. They started to take over the city and the government only to find that Hitler was still alive and vengeful. He ordered Himmler and his SS to round up not only the conspirators but anyone suspected of talking disloyally about the Führer. Himmler's men arrested over 7,000 people, including many senior army officers. Of those, 4,930 met their end in various unpleasant ways. One of them was General Fellgiebel, who was Wilhelm Flicke's commanding officer and a good friend as well. Fellgiebel was hung on a meat hook by the SS as they did with many others in a savage bloodletting. Germans in both high and low positions would remember the slaughter fearfully as the war came to its end.

There were other, larger casualties in the unreal world of the Third Reich as it collapsed. Allen Dulles, later head of the CIA in Switzerland, reported to President Roosevelt that: 'The period of the German Secret Service is drawing to an end'. He was right; Himmler and his lieutenants in the SS were triumphant at the discomfort of the Abwehr and its Director Admiral Canaris. The man who wanted to destroy him in particular was General Walter Schellenberg of the SD and his boss Ernst Kaltenbrunner, who was second only to Himmler. They busied themselves in rearranging their security organisations and the responsibilities for running them as the stricken German ship of state sank, but they needed a weapon to strike Canaris the final blow. This came in the unlikely shape of 'Frau Solf's Tea Party'. The lady was the widow of a former German ambassador to Japan and an anti-Nazi who was known for her formal tea parties, one of which she held at the end of September in 1943 for a number of her 'safe' friends. One that was not so safe was a young Swiss doctor who listened to the bitter anti-Nazi chatter who reported it all back immediately to the Gestapo, resulting in a whole swathe of arrests. Panic spread from Berlin and across Europe as even those who had only a passing acquaintance with Frau Solf's group were in danger. The bad news soon reached Istanbul where Erich Vermehren and his beautiful aristocratic wife, the former Countess von Plettenberg, were acting as senior Abwehr agents; they received orders to report back to Berlin to answer questions. They immediately telephoned the British Secret Service people at their embassy to beg for asylum and were whisked off to the security of Cairo. It was rumoured that they had carried the German Diplomatic Code keys away with them and many other secrets. Hitler was furious and raged against the disloyalty of the Abwehr, of which General Oster, a senior officer in the agency, was a known conspirator and they had also failed to alert the OKW about the landings of the Torch Operation or the Normandy landings. Now disaffected agents were leaving the sinking ship and the feeling was that the Abwehr was rotten to the core and its people were all traitors.

In February 1944 the Führer dissolved the Abwehr and handed its organisation over to the tender mercies of Schellenberg for the SD to integrate into the secret service to be run by the SS. The Abwehr had been operating fairly efficiently up to this time issuing chilling evaluations of the military and economic situation that were far too accurate for Hitler to read and almost treason for anyone with access to them to quote in public. Admiral Canaris, who headed the Abwehr, had been retired, which was surprising as the Gestapo had a thick file on his lack of loyalty to the Führer and the Nazi Party. That was generally a quick way into the concentration camps or worse. Canaris was put out to grass in a sinecure entitled the Office for Commercial and Economic Warfare and General Schellenberg achieved his ambition of becoming head of the Abwehr. He destroyed it and integrated the remains into his SD organisation, creating a vacuum in military intelligence as the SS produced intelligence evaluations that Hitler would want to read. This left the Wehrmacht, and particularly the army, blind for the rest of the war with no military evaluations on which they could depend. Meanwhile, Admiral Canaris experienced the inevitable knock on the door from the Gestapo and finished up in Flensburg concentration camp; after the obligatory amount of torture he was shot on 8 April 1945, twenty-nine days before the end of the war.

The War's End

Wilhelm Flicke recalls the disorganised chaos at the top and how it was reflected all the way down the organisation. The search for traitors and the discipline of terror gave the armed forces great solidarity with meetings of military men diligently and ostentatiously doing their duty. The war's end for Germany in May 1945 was cataclysmic as all community structure disintegrated. The author recalls seeing the way both young and old studied tattered pieces of paper pinned to notice boards outside ruined railway stations containing pleas to find members of their family with whom they had lost touch in the madness of war. One vivid memory of the time was a note of two children who had walked with their mother across half of Germany to the ruins of Berlin trying to find an aunt to help them. The mother had died at the roadside and the children could not find the street as the ruined streets and houses were unrecognisable and they were completely lost. Society had lost its cohesion and was only slowly beginning to regain a semblance of working order for its citizens in 1947. The Wehrmacht had ceased to exist as soldiers in tattered uniforms trudged home to ruined cities and often ruined homes and untraced families. Wehrmact and Abwehr personnel recounted how their units dissolved as staff were ordered to carry away purposeless files of orders and directives addressed to army or Luftwaffe units that no longer existed as the Allies advanced. Shattered offices on the Bendlerstrasse were emptied just before the Russian tanks rolled in to make the chaos complete. Secret intelligence evaluations were scattered and blowing in the wind and rain around the cobbled streets of Berlin.

German intelligence documents and signals transcripts found by the author lying scattered in puddles on the floors of concrete a blockhouse in Berlin's Spandau district contained the secret fruits of four years' signals intelligence work. The German signals intelligence services were destroyed or rather fell to pieces in such a chaotic manner that the story of the Abwehr has largely been lost. Accounts of some of those that had survived and served in the signals intelligence war along with some documents that did survive have been the motive to write this book.

Beginning Again

In the Second World War the American, British and Russian governments worked together to defeat Nazi Germany. They fought the Nazis at great cost to themselves until the war had ended and Germany surrendered in 1945. There always had been disagreements between the partners, although the overriding objective was to win the war. Now that objective had been achieved, each of the Allies was to act in their own interest, and the Soviet Union was to take a hard line in its demands, so that military action did not seem out of the question. This was a surprise to the other partners, particularly America who found herself unprepared to combat the Russian position and did not even have an appreciation of the Red Army order of battle, Russia's economic strength or the support that countries of Eastern Europe would give her in any kind of showdown. The Americans were given an opportunity to take over a major espionage network that was targeted against the Soviet Union, and this was to be one of the great intelligence coups in modern times.

General Reinhardt Gehlen had been in command of the OKH Army Supreme Command Foreign Armies in the east anti-Soviet espionage operations centre on Germany's Eastern Front. In January 1945 he was called to report to Hitler in what turned out to be his final visit to the Reich Chancellory. The negative response that Gehlen got from Hitler, who refused to accept the dire intelligence evaluation of the position on the Eastern Front, confirmed in the mind of the general the action he had to take. The German Army's resistance was about to collapse and the war's end was very close. Returning to his headquarters at Lossen near Berlin, he told his staff to be prepared to be evacuated before the Red Army arrived. He ordered the files in his centre be sorted into unimportant ones for burning and the important ones to be microfilmed three times and put into three special metal containers and sealed. As Russian tanks reached the Oder, he and his staff began their move east into the mountains in Bavaria where a Nazi Redoubt was planned to hold out until the last. Gehlen did not intend to resist but gave himself up to the Americans and was taken into custody to be interviewed by General Edwin Silbert of United States Army Intelligence. Silbert had an interest in how the Abwehr operated during the war, but Gehlen was not interested in history and was to make a proposal to the United States government about the future, asking more questions of Silbert than he as the interrogating officer was asking.

The American Office of Strategic Services was the main espionage agency of the US but had no intelligence sources within Soviet Russia even though relations between the two countries were beginning to be strained and of course Gehlen knew that. The contents of his boxes of microfilms contained details of a ready-made network of agents experienced in gathering intelligence about Russian military strength. He offered the Americans an evaluation of the Red Army's current order of battle and dispositions. General Silbert was given a quick overview of Gehlen's microfilmed Russian documents with descriptions of Soviet espionage organisations and some of their secrets. Within a week the German general was given an office in the US Intelligence Headquarters at Frankfurt am Main. He had to prepare a plan of action for redeveloping a new intelligence network within the Soviet Union's sphere of influence in Europe. American intelligence officers started coming to Gehlen with all sorts of problems. Documents printed out from his microfilms helped to validate his answers. That was not all, however. Nobody knew better than Gehlen that such intelligence would age very quickly and become dated. He began to reactivate the agents that were still in place to provide answers to pressing intelligence questions the Americans were asking about the current situations. Before long he had an impressive flow of information coming back from his network for evaluation that proved relevant to those urgent American intelligence needs. Washington was interested and Gehlen was soon invited there to discuss what was needed to form an intelligence service to operate behind the Iron Curtain. He made several conditions and the main ones were that Gehlen himself would direct the organisation and be in charge of his staff who would only report to him. He was adamant that he would not work against Germany's national interests; Gehlen had been anti-Nazi but very much a German patriot. This was true of Canaris and virtually all the conspirators involved in the bomb plot against Hitler. The final condition was that when German sovereignty was regained his organisation would be offered to the new German government.

The Americans agreed to it all and financed the new Gehlen Organisation with almost $4 million a year, which turned out to be a very good investment for them in the coming Cold War. Gehlen set up his headquarters in Munich to house his growing staff and started to recruit from a large pool of potential agents and ex-intelligence officers left over from the war. He even included ex-members of the former *SS Sicherheitsdienst* as the Gehlen Organisation built up its reputation in the international intelligence community. He pulled off a number of espionage coups, including planting an agent in the office of Ernst Wollweber, the head of the Communist espionage agency in East Germany. Gehlen's organisation was probably at its busiest before and during the Berlin Airlift when his intelligence covered almost every aspect of life in East Germany and the rest of Eastern Europe. The author first came across some members of his organisation during this time and many of their reminiscences appear in this book. Wilhelm Flicke worked for the Org, as its participants called it, under its Chief Cryptanalyst Dr Heutenhain in the SIGINT branch of the Gehlen Organisation. German intelligence took on a new lease of life as the recently formed CIA handed over the Org to the newly formed German Republic in 1948 with Gehlen still at its head. It retained its old staff and headquarters near Munich but was soon given a remit to expand its basis of investigation by Chancellor Konrad Adenauer. The agency expanded from its Cold War military base to one that gathered intelligence from every economic and technical aspect of life in Eastern Europe and further. The West German Federal Agency *Bundesnachrichtendienst*, or BND, is today acknowledged as a world class intelligence agency with a capability gained during the war to gather internal or foreign signals traffic and other intelligence using the most sophisticated espionage methods. Adenauer and all succeeding chancellors have expanded the brief of the agency for many purposes as an economic as well as military tool that helped build Germany's place in the world.

In Conclusion

The true effect and value of signals intelligence in war is difficult to calculate as the author well knows. Its combatants do not know how they have performed against their opponents until long after the effects of the signals intelligence war have died away. Bletchley Park's gigantic effort is said to have shortened the war by some years – but the evidence for that took over thirty years to begin to emerge. Contributions made by other intelligence agencies need to be at least outlined if the Electronic Order of Battle in the SIGINT struggle is to be properly appreciated. Signals operatives all over Europe, and indeed the world, strained every sinew to break enemy ciphers and codes during two wars and even in peace to gain an intelligence advantage. Some of those cryptographic operators began their apprenticeships in the Arendt Service of the Kaiser in the First World War and were still serving at the end of the second one under Hitler. Their length of service for the German Abwehr compared with that of Alastair Denniston's Bletchley Park team. The author got to know and befriended a group of these German Abwehr veterans as the war came to an end and spent much time listening to their reminiscences and reading the papers that they offered him. This gave the author a more rounded view of the various SIGINT activities of the Wehrmacht operators as they tried to surmise their enemy's intentions by evaluating the intelligence performance and actions of the Allies during those years of conflict.

Our clearest insight into German achievements came from the TICOM investigation that was carried out by the Allies immediately after the war. This was not really an investigation at all but a cherry picking of the intellectual property in the new weapons that had been developed by the Nazi scientists. The team of Allied cryptanalysts were actively seeking specific innovative equipment, codes and techniques that the Abwehr were about to launch as the war came to an end.

The innovations that were appearing in German signals intelligence were too late in the day, as was the case with most other weapons development at the time; it was simply all too late to influence the final outcome. This was due to another of Hitler's errors of judgement, after the Battle of France he ordered that all technical development with a lead time of more than two years be discontinued as he expected only a short war. Some of the new and largely untried equipment and their techniques were very sophisticated and could have given the German signals intelligence agencies a huge advantage over the Allies, but it was not to be. Much of the 'stuff' developed was snapped up after the war by either the Russians or the Allies, in any case it all disappeared behind a veil of secrecy to serve in the Cold War that was beginning to emerge. But that is another story.

The story of German signals intelligence in wartime, and even its earlier history, emerged as the author talked at length (or listened to) veteran operators about their military intelligence experiences in the late war. As ex-Wehrmacht signals veterans they were convinced that they were at least deserving of equal honours with their Allied compatriots in the signals intelligence war – a stance that seemed logical and reasonable to me at that time. Indeed, as they swapped war stories, most of them were sure they had been the victors in the war in the ether – the fault, they thought, was a lack of force and opportunities available to take advantage of that knowledge. That was certainly true in the latter stages of the war, but also by then the domination of Allied intelligence in the shape of Ultra and Enigma began to emerge. Those decreasing number of German veterans I had been able to keep in touch with looked on in disbelief as the story of British cryptography unfolded. Today the superiority of the mobilisation of the British intelligence community and its dominance over Germany and her allies is undoubted, but there was little evidence of that at that time. A series of books recounting Bletchley Park's story began to be published which shattered the conviction and confidence of these veterans in their own signals intelligence performance. The huge intelligence advantage that Bletchley Park held and how badly their service performed compared with the Allies and the deceptions they practised on the German High Command (who wanted to be deceived anyway) emerged only gradually. Reviewing the TICOM 'investigation' and other evidence now available (some documents are still withheld) shows that Germany did as much to lose the signals intelligence conflict as Britain and the Allies did to win it.

As the truth dawned on my Wehrmacht friends I well understood their conviction at that time that Enigma and the related codes and ciphers had remained unbroken. I believed their assertions of competence and even superiority in their skills in operating their machines and communications systems security, as they were made with such certainty and had a confident ring of truth. The German High Command's received wisdom that their codes were unbroken would have led them to believe that their whole communications system was unpenetrated; it would have taken a questioning mind to doubt it, although some did in private. It would also have taken courage to suggest that their strategic communications were flawed, particularly as the price of failure in Nazi Germany was often high. The decreasing number of aging German veterans that I still kept up with over the years looked for solace in the fact that interception and evaluation of enemy transmissions at a tactical level by German operators was of good quality. They were right about that, signals intelligence in frontline units from army group down in the military hierarchy was of a high quality right up to the end of the war. Principal among these signals intelligence warriors was Wilhelm Flicke who became quite a close friend and from whose papers I have quoted liberally in this book. The alternative story about Colonel Bonner Feller and the American Army, and how the 'leak' was discovered by a radio broadcast came from Wilhelm. The different versions of how coded reports of how British Army dispositions being sent back to Washington were broken, unaware that Rommel was listening, are also significant. The reason why the Allies won the signals intelligence war was that they were more convincing liars than their opponents; that is why there is never just one truth in intelligence operations but many.

Another of my German veterans, Wolfgang Schrader, was asked what went wrong with his system which led to a discussion about the rights and wrongs of intelligence in general. He gave four rules for an effective working military intelligence unit: trust, cooperation, organisation and objectives.

Undoubtedly trust is needed, not only in evaluations of enemy actions and objectives but also in the competence of the organisation and even the concept of intelligence itself. The confidence of the leadership was important and certainly Hitler would have been a more better military leader if he had managed his intelligence community more effectively. Hitler did not so much distrust his agencies as discount them because they did not share his political allegiances. The German intelligence community had a foot in two camps, the German patriot and the Nazi fanatic, and this was the besetting sin of German intelligence and indeed to some extent the German Wehrmacht. As we have seen, the Abwehr and the SD were constantly at each other's throats because Hitler had allowed and even encouraged them to be, so the feud between them was particularly corrosive. With no trust there could be no cooperation and the collation of intelligence evaluations requires a cooperative effort. Lack of trust and even cooperation seems to have been a common pattern in all agencies in the intelligence community around the world, which lasts to some extent even to this day. The American army and navy were as spiteful to each other as the German Abwehr and Gestapo were, although the Americans stopped short of killing each other. Japanese military and naval cryptographers did not speak to each other either. The British prided themselves at being the best operators in the field of intelligence but even their Secret Intelligence Service and Special Operations Executive were not above a bit of back-biting. It seems that the closer those organisations were to each other the more they bickered, if you can call the Gestapo hanging their opponents on meat hooks bickering.

Trust came mainly from the top; Churchill had his faults in strategic wishful thinking in the early stages of the war but was gradually schooled by his senior staff, particularly General Sir Alan Brooke, to trust and listen to what intelligence evaluations had to say. Stalin refused to listen to Churchill's intelligence-based warnings indicating that Germany was about to attack Russia – he had to learn the hard way.

Allies certainly need to trust each other and agree intelligence objectives, and one of Germany's major problems was the waywardness of her friends. The Abwehr should have been watching them more closely but German intelligence did not work very well on an international level, particularly as the war progressed. Their opponents had their differences of course; for instance the Americans did not like way the British ran their empire but militarily their objectives were generally in accord – not so Germany's allies, Italy and Japan. Italy wanted to expand her empire but did not have the military resource or will to do so. When Mussolini's army failed in its tasks in the Balkans, Hitler was embarrassed into military adventures there and in North Africa. The German army had to take over the mess on each occasion which skewed the strategic plans of the German High Command. Inevitably, they finished up fighting, not only on an Eastern and Western fronts in Europe but on an unplanned southern flank as well. If the Italians were an embarrassment to Hitler, her Japanese partner who had been a signatory to the Tripartite Pact in 1940 was even more so. The pact contained promises to help each other militarily including cooperation in cryptographic matters. During the autumn of 1940, Hitler realised that the Japanese were not going to attack Russia. The Japanese diplomatic code had been broken by the Abwehr, according to Wilhelm Flicke Tokyo's instructions to their ambassador in Berlin about entering the war were being intercepted. Obtaining the American and British ciphers that the Abwehr had broken was very important to the Japanese in attacking their targets in the Pacific but they also insisted that Germany declare war on America. One of the great mysteries of the Second World War was why Hitler agreed to it. If he had refused, then President Roosevelt would have found it politically impossible to involve his country in Europe's war, but he did.

Japan attacked Pearl Harbor on 7 December 1941 and Hitler unwisely kept his promise to make one of the war's most important decisions and declared war on America on 11 December. German cryptologists began cooperating with Japan immediately but it seems to have proved a fruitless exercise. Hitler's decision was difficult to justify but then so were many of his often ill-considered and impetuous orders.

One assumes that the Führer felt that, if he could not persuade the Japanese to attack Russia then an attack on America would be the next best thing as they were inevitably drifting towards war with Germany anyway. If Japan could keep American attentions fixed on the Pacific, so that they did not interfere in Europe, it would make things easier for Germany. He was wrong, Churchill and Roosevelt agreed that Germany's defeat in Europe would be their prime objective and only after that would the Pacific get the full attention of America's fury. Thus, their objectives were aligned in a way that the Germans had not envisaged.

Wolfgang's other condition on the list was the organisation of intercepted signals traffic prior to turning it into good intelligence evaluations. The secure and orderly management of huge flows of intercepted raw data and its decryption is an essential part of the modern signals intelligence process. The creation and maintenance of a database of thousands of discrete pieces of information that can be turned into usable intelligence evaluations is a major task. The job today is done by computers but during both world wars the recording of all this information would be done in pen and ink on index cards. The Germans are generally thought of as an orderly and organised people but they made a dog's dinner of their intelligence records and the way that they were organised. Hitler's disinterested management attitude to intelligence may have had something to do with it, but the Abwehr and SD deciding to remain in central Berlin until their offices and databases were bombed to destruction did not help.

The Germans never discovered the secret of concentrating all their best cryptographic people into one location away from the bombers. The turf war in German intelligence did not allow their gifted intellects (of which there were many) to interact with and nourish each other in one place, unlike the British organisational inspiration for their counterparts.

Beyond Schrader's criteria, luck is the additional magic ingredient in all SIGINT projects, in common with virtually all other military success. Leadership and discipline have always been important to military operations but good fortune has similarly been a critical ingredient. Napoleon always asked any of the commanders that he was going appoint 'Are you lucky?' Good luck proved to be the most effective way to break through the codes or ciphers of the enemy and the British intelligence community has always had an incredible amount of that invaluable commodity. Within days of the beginning of the First World War, some accidentally intercepted German wireless signals were passed to a senior naval officer who fortunately knew the right man to deal with them. Sir Alfred Ewing began to lay the foundations of Britain's Room 40 intelligence agency at the Admiralty, an organisation that had an even more profound effect on the outcome of the First World War than Bletchley Park had on the Second. His headstart in obtaining, one by one, the three German *Hochseeflotte*'s code books to make a full set was the luck of the devil. Building on that run of incredibly good fortune he and his team created their intelligence agency through hard work, and detailed and dedicated organisation. Meanwhile on the other side of Europe, the German General Paul von Hindenburg, who was about to fight a battle with the Russians at Tannenburg, had his own piece of good luck. Operators in a nearby radio station intercepted Russian messages in clear language detailing the Imperial Army's plans and enabling the general to win a resounding victory that changed the shape of the First World War. Where Hindenburg's story differed from Room 40's was that he did not expand the opportunity that chance had given him. He did not create an organisation to deal with the unencoded (or encoded) messages that he was to encounter or train his operators in the security measures that he needed to fight against the French *Deuxieme Bureau*. It would be a year before the German Army had a working intercept agency of its own in France. The Battle of the Marne might have gone differently if the Germans had a trained a cadre of signals operators in the use of a secure system. Each of these occurrences was built on an error in security by the enemy – the signals intelligence war is not forgiving.

The decade between the two wars saw the development of signals technology and the design of an automated ciphering process mainly using a plethora of coding machines. A small band of Polish cryptographers showed the intelligence community that such machines were not always secure and could be decrypted by dint of great patience and ingenuity. With huge generosity the code breakers of Poland handed the machine to the small band of ex-Room 40 cryptographers with notes on the work they had done to understand its mechanism. It was a huge piece of luck; the world and particularly the British owe the Poles a great deal. The war did not begin well for the British but as the tide turned the golden age of signals intelligence began. The concept was born in the British army and navy, and soon spread to the American forces; like most things the British invent the Americans took it over and turned SIGINT into the global power it is today. The Germans thought they were doing well in their ability to read other people's military mail but it became obvious that they could not compare with the deviousness of the British. Although undoubtedly, the intelligence skill that they had acquired in the first war survived through the inter-war years to emerge in the Second World War as a powerful weapon of war.

The survival of intelligence agencies and their skills after times of need in war is a thread that runs through the intelligence story. In times of peace and economy politicians have to decide between the cost of investing in expensive and effective weaponry compared with the cost of not having it when it is needed. It has always been a tough call but the skills of intelligence management seem to have had the habit of surviving in some form in the armies of most major military states.

Half a century of Morse-coded ciphers and decrypts came to an end, or rather took a different shape as the BND came on the scene. Code breaking operations in the Buckinghamshire manor house came to a fairly abrupt halt at the end of the war but not before the stirrings of its successor at Government Communications Head Quarters (GCHQ) took up the torch in the signals intelligence war. The Cold War was to prove a different form of conflict with a different enemy and with electronic weapons whose nature had changed. The computer replaced the dependence on index cards and records written in pen and ink and even more, it recognised complex patterns and made connections that would have made the staff of Bletchley Park, let alone Room 40, deeply envious. Alliances between agencies are more common and include the BND as an important partner to the CIA, SIS and others, some of are not publicly acknowledged.

The widespread and changing aspects of intelligence emanates from agencies that still have their roots in Marconi's wireless transmissions of Morse code. The dots and dashes that filled the ether for a few decades have now faded and passed into history, giving way to more sophisticated signals serving differing purposes. Yet, the contribution that the world's signals intelligence community has made to fifty years of war-torn history still needs to be defined. I hope this book provides the beginning of a framework for the telling of this story.

Appendix

This is a transcript of a Top Secret document from the United States government's archives from a copy held by the author. The typewritten thirteen-page document is so faded that it is not possible to reproduce the original so the content has been reproduced and the front page is overleaf.

25 copies of which this is No. 13
40/48/TOPSEC/AS-14-TICOM

Sabotage at the German Intercept Station at Lauf

1. The attached is an Army Security Agency translation of TICOM Document IF-294-3. It is a paper written by Wilhelm FLICKE, formally a member of the LAUF intercept station. German title 'Fall Lauf'.

2. This is one of a series of papers written by FLICKE dealing with his experience at the LAUF intercept station of the Signal Intelligence Agency of the Supreme Command German Armed Forces (OKW/Chi). They were received by Army Security Agency through non-TEACOM [sic] channels and no check of the veracity of FLICKE story is possible for specific details. In general, however, FLICKE appears to be a remarkably well informed source of information on both OKW/CHI at LAUF intercept station, and the appreciation of signals intelligence by the German leaders.

3. 'Der Fell Lauf' is an interesting account of sabotage and as such merits the attention of those charged with counter-espionage as well as of intercept units. As most of the persons mentioned by FLICKE escaped death at the hands of the Gestapo and are presumably still active in

clandestine activities FLICKE used cover names throughout his narrative. The identification of the real people behind the cover names has been included on the last page of the document.

IF 294-3

The 'Lauf' case

If one were to try, one would search in vain for great bundles of documents on this affair. There was no monster trial; all participants are still alive and at large. Their names therefore, will be changed in this narrative. The whole case was never uncovered in all its details and ramifications by German officialdom. Like the case of the 'ROTE DREI', it remained known to only the small circle of participants. Unlike the case of the 'ROTE KAPPELE', these people did not get into the whirlpool of engulfing fate.

The comparison with the cases 'ROTE DREI' and ROTE KAPPELE' is not accidental. For in its practical results the 'Laufe Case' can take its place along with others, in a quite different form.

But while in the case of the 'ROTE KAPPELE' it was a question of a widely ramified resistance organisation and a large number of important personalities collaborated, the resistance group of which we shall speak here consisted of only a few persons in very subordinate official positions who otherwise played no important role in life.

Reference has been made elsewhere to the fact that the high command of the armed forces had two intercept stations for monitoring all international, diplomatic and military-political traffic. Those stations were set up and put into operation in 1939. The technical equipment of those two stations of which one was in Treuenbrietzen near Berlin, the other in Lauf near Eihenberg, was thoroughly up to date, the organisation was as perfect as possible. Since the assignment of these stations was principally concerned with intercepting well known high-powered stations which worked at [*text too faint to read*] lengths, any grouping in the [*text too faint to read*] necessary to tune in on definition [*text too faint to read*] the transmission would be heard.

Most of these messages were sent but by so called high-speed transmitters (SCHNELL sender). [The rest of this passage is too faint to read properly but seems to be concerned with the kind of transmitters that are in the LAUF intercept station.]

Whereas in the intercept services of the Army, Navy and Air force, the work of the intercept operators was to some extent individual and the man at the receiver had a certain live contact both with the station to be covered and ultimately the analyst. In LAUF and Treuenbrietzen a system of division of labour had been created in order to get the highest possible quantative results from the operator.

When in 1942 reductions were taken everywhere in the offices of the German armed forces in home territory, this man-saving Taylor system grew from month to month. From an individual working method there came a more and more impersonal, factory-like procedure, so the two intercept stations, each of which had been eager to appear better and more efficient, gradually in particular the station at LAUF took the character of radiogramme factories.

Early in 1943 it was decided to replace part of the mail crew by feminine assistance (so called NACHRICHTERHELFORINNEN), in order to free men for the needs of the front.

In February 1943, the first female command of this kind arrived in LAUF. There were some 25 young girls who had already been trained as operators and were now given a special course in preparation for their special assignments. Among them was a blond girl some 19 years old, loaded with sex appeal and other qualities who did not however stand out particularly by virtue of her work and who seemed at first to take no very great interest in her duties or to stand mentally above average. She claimed descent from French nobility, which was backed up by the 'de' in her name. The young lady's father had a postage stamp business in Western Germany and supposedly had intimate relations with people abroad before the war.

Miss de Villiers was trained therefore along with the other young ladies and after about six weeks took up her duties as one of the 'receiver groups'.

It is uncertain when the English Intelligence Services first made contact with her. Probably it was through her father. In any case, from April or May 1943 on she had regular meeting with a British Agency in Nurnberg in the narrow alley behind the Grand Hotel opposite the main station and the UFA_KINO. The spot was well chosen because many loving couples met there in the darkness before and after the movies.

The first thing Miss de Villiers did after she came to be at home in her job, was to remove and turn over to the agent gradually a sample of all the amplifier tubes used in the apparatus. With such large scale operations, this did not attract attention. She also gave him all desired information regarding operations and the persons engaged in the intercept station.

This in itself was no unusual piece of espionage work and rarely called for some skill in the doing. Much more interesting was her next step.

With the end of 1942, America had entered the war. At that time interception of American diplomatic and military political traffic, was the focus of interest for the German Intercept and CRYPTANALYTIC service. Especially important was the copying of all traffic between the Secretary of State in Washington and the representative of the USA abroad. A cryptographic system employed for that purpose was being used in 1944 in Berlin making it possible to read large parts of messages between Washington and representatives abroad. Radio traffic between the United States on the one hand and Switzerland, Turkey and Sweden on the other was of primary interest.

The messages were sent in five letter groups and bore in their heading the word 'Socstate', after going to Washington; then directed abroad they had as address 'Amlegation', 'Amembassy', 'Amleg' and generally the signature 'Hull'. These were most important communications. Hence interception of these transmissions was given top priority. In the instruction to receiver operators, interception of these messages topped the list.

Since the Hull telegrammes were very important, they were sent out from America by two linked transmitters on different wavelengths so as to assure proper reception under all circumstances. The type of hook-up depended on the season, the distance to the receiving station and the receiving conditions at the moment.

In the main, stations HQK, HQK-2, HQG, NEG, NEG2, ECI, NDE, and LDU came into question. All told some eighty stations were observed.

Since the Germans attached the greatest importance to complete coverage of the traffic (justified by Cryptanalytic considerations alone), the order was given to monitor both wavelengths and have two different receiver groups do this.

From this fact, every receiver operator could judge even without being told, that these telegrams were being deciphered successfully in Berlin. It was equally well known to all operators in LAUF that decipherment of Russian five-letter diplomatic traffic appeared hopeless, since its interception was not required. The plan which was now put into execution was heard in these circumstances.

[Missing text, too faint to read] in contact to sabotage interception of a group's radio circuits; among them was the 'Secstate' [text too faint to read], and amleg telegrams. This could not be done without more ado, since the double monitoring of this traffic made it impossible to suppress intercepts. Hence a different [text too faint to read] had to be taken.

Making use of feminine charms and capabilities, Miss de Villiers built up a group of conspirators within the Lauf intercept station to which belonged primarily those receiver operators [text too faint to read] in intercepting the 'Secstate' and related radiogrammes. These were the [text too faint to read] orders (In civilian life Goldsmith in Nurnberg). Waldmann [text too faint to read] had previously held a high position in German diplomatic service and the operators Zerbst and Droppe, regarding whom she knew they had communist meanings. The two first [text too faint to read] were strongly anti-political and anti-communist. The plan was to permit no outward change in the reception of Secstate messages but in reality to stop messages but in reality to stop decipherment or at least to interfere with it. Therefore the telegrams continue to be intercepted. At the same time however, on another high speed machine, five letter [text too faint to read] diplomatic radio telegrams were intercepted, the home tape of the Secstate telegrams was cut off after the heading and the indicator group, a corresponding number of Russian five letter groups were pasted on, followed by the Secstate dispatch. The first cipher group was the sol called indicator or key group, it differed with the address. Only by its aid could the true recipient decipher the messages. In preparing the telegrams, the indicator was left untouched, frequently several groups were also left after the indicator.

The genuine Secstate groups were then thrown in the wastebasket. The doctored tape was carried to the reading room and there treated as described above. Naturally, decipherment of such telegrams was no longer possible in Berlin.

In introducing the Russia Cipher groups, it was necessary to be very cautious as the Russians use a type of five-letter groups which were put together in a particular fashion. Of course, these would have been immediately attracted to the attention of the analyst. It was necessary therefore to pick out such cipher text as differed in no outward respect from the American traffic. To avoid any breakdown, the conspirators chose the following way. They arranged matters so that one of them was always in the booth engaged in interception of Secstate messages. As soon as the transmitter in question began to function, the conspirators told the operator making the second recording that he need not bother with that station because the conspirator was getting it perfectly. In this way they avoided having the message reach the readers and so eventually Berlin, in two different forms. The operator was relieved of this work in this 'friendly' fashion was not always enthusiastic about the favour, for his production curve sank in consequence. Regarding this curve a word of explanation may be in place. The Director of the station, a Major X, who was in civil life a manufacturer in Saxony, had applied to the intercept station in Laugh the Stachancw [*text too faint to read*] methods of his enterprise. He prepared performance record cards for each employee as they served on the basis for monthly graphic performance curves.

Flicke's Note:

Let us assume that 'Secstate' messages were received in booth VIII or IX while any innocuous operator or a Nechrichteshelferin provided the parallel coverage in another booth along with other assignments.

The performance of an operator was computed on the number of cipher groups intercepted, especially those of the first rank so-called, i.e. of important telegrams which either could be deciphered or where there was prospect of decipherment. Hence the operator had to try to receive during the month as many and as long messages of the first rank as possible. If an operator wanted to avoid being transferred to a front unit 'to improve his rations', he had to try to keep his production curve from sinking during the month and if possible keep it sloping slightly upward.

This applied also to the conspirators. Consequently they were among those who endeavoured to make a good impression on the chief by their performance. Of course, this could not be done at the cost of other operators; one good turn of the type described above meant inevitably a disadvantage for one's comrade in the noble competition for a place in the production honour list. To be quite safe in this respect, the conspirators capped the climax in their friendliness; they intercepted Secstate messages diligently, prepared them and gave the finished Morse tapes to their colleague on duty in the other booth who was happy to be able to improve his day's credit for intercepted cipher groups without any effort on his part. Miss de Villiers and her four co-conspirators were well liked among their comrades and got many a cigarette in return. Even with the director of the station, they enjoyed the highest esteem as good operators.

By this system the conspirators fixed it so that in the records, not only their names appeared as intercept operators for Secstate messages but also many others. First was a guarantee against suspicion. Now and then of course, genuine Secstate messages were intercepted and forwarded. At the cryptographic section in Berlin, terrible confusion resulted. At first it was thought there had been a faulty transmission by teleprinter. However, comparison with the forms filled out by hand which arrived on second day following showed that this was not the case. Then it was assumed there had been errors by the reader. They called up Lauf and requested a check but even this idea soon had to be abandoned since the Morse tapes which were preserved for several weeks in Lauf proved upon checking with the forms sent to Berlin that the readers had functioned correctly. Now the suspicion arose that the operator during reception that night had accidentally gotten off on the transmission of some other station; in that case the Morse tapes would have had to show shifts.

Again, they called up Lauf. Again they checked but nothing of the kind was found. No intercepts could be more painstaking and accurate. The operators, readers, teleprinters which they had been ready to cuss out, ultimately were praised. The proof that the mistake was not at the intercept station at Lauf was found in the fact that this remarkable phenomenon did not occur in the work of some inexperienced girl operator but gradually was found in the work of all operators serving the receivers in question.

The cuts in the tapes could not arouse suspicion since for example the heading of the telegrams were regularly re-pasted in transmission or messages were frequently sent in parts which other unrelated telegrams were sandwiched between either transmitting station. Hence such messages were cut out to lighten the work of the readers.

In Berlin they tore their hair. It was clear that the Americans had changed their cipher or were using two different systems side by side in the same traffic. They were still glad they could decipher at least a part of the informative radiogramme. The Secstate messages continued to rate top priority. And the operators at Lauf stepped up their production curve so that

Major […] rejoiced. These capable youngsters must be kept at the station; it was impossible to think of letting them go to the front; but the work of the five conspirators was not exhausted in the handling of the Secstate messages. There were other fields of activity. In the Spring of 1943 the war started in North Africa. In the summer, operations extended over to Sicily and Italy. General de Gaulle developed great activity in creating a new French armed force. His 'Francelib' radiogrammes went all over the world, especially from North Africa to Casablanca, Madagascar and London. These were five digit groups and five letter groups. The one set contained important information and for interception rated as telegrams of the first rank; others dealt with less important matters and ruled as telegrams of the third rank. The conspirators treated the message by intercepting them all and then cutting out the cipher text of important messages and inserting more cipher groups of this less important; the text they had cut out wandered into the waste basket. In this way more than half the intercepts of the 'Francelib' messages, in particular the Francelib–London, were sabotaged.

Flicke's Note:

The interception of Francelib messages was later assigned to another intercept station since Lauf proved unfavourable. People in Berlin never did get wise to the real situation.

In this last field the five conspirators found another ally in an operator Ley, who came from the West German frontier area and had relatives in Alsace. Apparently, he had hit on the idea of sabotage all by himself and had confidential relations only with operator Anders. For the rest he worked independently. On the other hand the remaining conspirators who operated very closely in the doctoring of their messages.

Such a change in message texts was possible only because the analysts in Berlin had no idea of the practical set up of the work of the intercept station. In earlier years the head of the Berlin evaluation section had suggested to the head of the cryptanalytic section again and again the need for close co-operation of all organs and persons engaged in the intercept services, but he had always been rebuffed brusquely. The analysts therefore had a secluded existence in their own rooms and could do nothing except juggle digits, groups of letters and statistics. If one of them had taken the pains to go to Lauf and subject the Morse tapes to a critical study, he could not help noticing that ahead of all the undecipherable texts Secstate, Amembassy and Amlog messages or of important Francelib text there was regularly a cut in the Morse tape which could not always be explained as harmless.

Our nineteen year old blonde, de Villiers, was and remained the head of the group. This was all the more astounding since Major X was unceasingly on the watch for traces of illicit love affairs among his helpers. Striking up and maintaining such liaisons had therefore become a sort of sport at the Lauf intercept station. It was a matter of honour to play a trick on the chief in this respect. Miss de Villiers always carried in her handbag a loaded L'Uicue pistol, calibre 7.65. She was a crack shot with the pistol. Among other things she delivered currently to her employer in Nurnberg the lists of telegrammed addresses which the control station of the intercept service in Berlin issued to guide operations on the interest stations from those the English Intelligence Service could see which systems were deciphered in Germany and which were not.

In 1942 at the intercept station Lauf Treuenbrietzen a change had been made from linear antenna to directed rhombus-system for receiving certain stations which were difficult to hear. Huge structures were erected at great expense in effort and material and soon a forest of masts surrounded the station buildings. The rhombus antennas pointed in the most varied directions and actually did improve reception. Their leads ended in a great switchboard in the central intercept office. From here there were leads to the various booths. Over the seats of the operators were frames with contact jacks. By using big thick metal plugs, each operator could contact each of his machines with any rhombus he chose.

He delivered these to the agent in summer 1944 and he passed them along to have a special device built in; this was to result in a shorting of the antenna and reduction in the possibilities of reception. It was the intention to rebuild all these rhombus plugs in the course of time; however, this plan was not executed in full.

In August 1943 Miss de Villiers married a soldier of the Tank Corps who in civilian life was a farmer in Saxony. The wedding occurred in Dresden and Miss de Villiers was given three weeks' leave. Before her departure she agreed with Anders that he should send her a telegram from Nurnberg on the day of the wedding with an urgent recall to the job. This was done and Mrs Kremer, the former de Villiers, spent three weeks honeymoon with Anders in Nurnberg where she had three or four meetings with her employer; new plans were sketched for sabotage at Lauf. One of these was very soon put into execution.

In the meantime the intercept stations at Lauf and Treuenbrietzen had each received a radio transmitter. It was intended for use in ordering the other stations or the secondary station situated in Lorrach to take over in case of thunderstorms or air alarms until the disturbance had passed. The transmitter in Lauf was connected to a linear antenna the ground switch of which was located outside a window of the service building. By slipping a metal ring between the ball of the antenna and the ball of the ground it was possible to ground and the transmitter could no longer be heard in Treuenbrietzen or Lorrach. Of course this procedure could only be used in darkness but it was employed repeatedly without anyone's becoming aware of the real situation.

Flicke's Note:
Such a system of sabotage could only be used when a young inexperienced operator was on the transmitter. Mrs Kremer extended this sabotage to the receiver by grounding its antenna and preventing reception of signals from Treuenbrietzen.

Late in July the German Secret Police arrested a man in the vicinity of Cologne who turned out to be an agent of the British Intelligence Service. From material found on him his connection with Mrs Kremer alias de Villiers was established. Accordingly her arrest was ordered by Cologne.

In a way that would take too long to describe here, Mrs Kremer was warned. She got the pre-arranged cue by phone. She immediately informed her friend and fellow conspirator Anders who immediately gave her the keys to his lodgings in Nurnburg. His comrade Waldmann hastened to Lauf railroad station and bought a ticket for Sulzbach-Rosenberg, but got onto a train for Nurnberg to thow pursuers off the scent. In Nurnberg she stayed in Ander's apartment and on the following day he provided her with a black wig. He also procured the necessary food. Meanwhile the criminal police zealously but vainly followed the trail leading to Sulzbach-Rosenberg.

Some months earlier Zerbst, against whom there was never the slightest suspicion had become very nervous for some reason or other and since at that time a capable operator was being sought by the German intercept station in Sophia, he applied for the post and left the next day via Berlin, accompanied by the best wishes of his Chief. In Sophia he immediately made connections with Russian agents and for a long time provided the Russian information service with reports. But when one day he thought that he was being watched he fled and went over to a partisan battalion where he soon assumed the leadership of a 'Zerbst group'. Luck would have it however that the traffic of this partisan group was intercepted and deciphered by a station of the *Funkabwehr*. At Zinna where the headquarters of the *Funkabwehr* was located and where the evaluation reports were composed, they naturally had no notion of the connection of the name Zerbst with the German Intercept Services. When the monthly report of the *Funkabwehr* arrived in Lauf early in September, people were very much surprised to find this familiar but not too common name in an entirely new connection.

At the Lauf intercept station at that time there was an agent of the Gestapo in the guise of Corporal. This man, we will call him Bohn, was outwardly an insignificant figure. Secretly (without the knowledge of the director) he was watching all members of the station and what went on there. Bohn had learned of the close relations between Mrs Kremer and operators Anders and Waldmann long ago without being able to establish any further significant details. He knew that Anders had been friendly with Zerbst. He now directed the attention of the secret police and the director of the station to Anders and Waldmann who were now quizzed searchingly with Major X; since the two knew the erotic complex of their chief they attributed the whole matter to the whole operation of their love affairs.

Meanwhile, however, Bohn uncovered a series of incriminating factors, above all detail of the formally close contact between Zerbst and Anders, and so the latter was arrested one day and brought to Nurnberg for a hearing. The hearing developed nothing tangible and ended in ordering a search of Anders' apartment in Nurnberg.

Anders heard this order with indifference. Taken back to Lauf, he succeeded in informing his comrade Waldmann who immediately hastened to Nurnberg and warned Mrs Kremer. An hour before the arrival of the police, Mrs Kremer left the endangered apartment and was taken by Waldmann to the goldsmith's shop of Anders from which she moved next day to his mother who likewise lived in Nurnberg.

Since the search of his dwelling was without result, Anders was again released. On the other hand Waldmann was arrested but he was also released after a few days since his father put up a large sum as bail. Both maintained contact with Mrs Kremer, using all due caution, but the way things lay it was merely a question of time when the criminal police would discover her trail.

Meanwhile in Lauf they had gone through the effects which Mrs Kremer left behind in her hasty flight. It came out that in the pockets of her grey linen coat which she used to wear for protection on duty there were several Morse tapes with the heading Secstate, Amembassy: the intervening text was missing.

After racking their brains as to why Mrs Kremer could have kept these messages in her pocket, they questioned Anders about it. With an unconcerned smile he said Mrs Kaufmann (Does he mean Kremer?) had always wanted to turn in the neatest intercepts; if now and then the beginning or end were somewhat bad, she simply took clean ones out of her pocket and pasted them.

Such an explanation appeared plausible and was received by the director of the station as well as his assistant in charge of intercepts with satisfaction and inner contentment; after all it showed what an interest the people displayed. They were convinced that the whole Kremer affair would turn out to be harmless.

After about three weeks in hiding, Mrs Kremer lost her nerve and turned up one morning in her black wig in Lauf where she lay in wait for Anders on his way to work. She declared she could not stand the nervous strain any longer and intended to report to the director of the station.

Anders knew from his own hearings the Gestapo had no real information regarding Mrs Kremer's activity but that the warrant for her arrest rested solely on suspicion; therefore he said that with the plan. You can imagine the astonishment of Major X when the two suddenly reported to him without warning. Mrs Kremer was taken to the Nurnberg police headquarters while Anders continued to go free.

Meanwhile the sabotage game had gone on merrily even without Mrs Kremer. However, it was to be feared that, although she refused to make any statement in spite of beatings and other inquisitorial methods of the Gestapo, she might someday make incriminating statements. Cleverly Anders and Waldmann found out the cell at police headquarters in which Mr Kremer was held. However, three attempts to free her failed and only led to Anders compromising himself and being arrested again. They found out that Anders had given her a hiding place; he was brought before a military court and charged with being accessory to flight and was finally transferred to a penal battalion in Norway where he soon made contact with the resistance movement. His real role in the group of conspirators at Lauf was not discovered.

Indeed nothing was discovered at all. No suspicion fell on Dropps. He was transferred in October with excellent recommendation from his chief as a good operator and good comrade to the replacement battalion in Eschwege and from there was assigned to a field unit. Waldmann was able to turn the whole matter to the erotic side. Lay was transferred in the summer of 1944 to another post and Mrs Kremer was silent. No threats or beating to induce her to speak. The Gestapo however was much concerned to find out the man in the back of her and any other conspirators. She was dragged from hearing to hearing until the Americans occupied Nurnberg and the Lauf case was settled automatically. In Berlin they were often amazed that the diplomatic ciphers of foreign countries were changed just at the moment that the Cryptographic section had reached its goal. They frequently believed in clairvoyance but there is no clairvoyance, only realities.

Author's note:

This wordy statement from a TICOM officer taken from Wilhelm Flicke when captured and interrogated is extraordinary. Probably the only way that Flicke can have known about this matter in such detail was if he was implicated in some way in the sabotage attempt himself. He was a known anti-Nazi and had been in trouble with the Gestapo which may help to explain the matter. The successes and failures of the German signals intelligence service in the war makes dismal reading but out of the ashes a Phoenix was to grow.

Bibliography

Andrew, Christopher, *Secret Service* (London, Sceptre, 1986)

Andriessen, J.H.J., *WW1 in Photos* (London, Robo, 2006)

Bennett, Ralph, *Behind the Battle* (London, Pimlico, 1994)

Buchheit, G., *Der Deutsche Gehheimdienst* (place, Econ Verlag, 1956)

Dourlien, P., *Inside North Pole* (London, Kimber, 1956)

Glantz, David M., *Soviet Military Intelligence in War* (London, Routledge, 2013)

Hagen, W., *Der Geheime Front* (Zurich, Weiner Verlagshaus, 1967)

Jackson, John, *Code Wars* (Barnsley, Pen & Sword, 2011)

Jones, R.V., *Most Secret War* (London, Hodder & Stoughton, 2009)

Kahn, D., *The Code Breakers* (London, Macmillan, 1963)

Keegan, John, *Intelligence in War* (London, Pimlico, 2003)

Kouzminov, Alexander, *Biological Espionage* (London, Greenhill Books, 2005)

Lewin, Ronald, *Ultra Goes to War* (London, Book Associates, 1978)

Mueller, Michael, *Canaris* (London, Chatham Publishing, 2007)

Paine, Lauren, *The Abwehr* (London, Robert Hale, 1984)

Simpkins, Peter, *WW1 1914–1918* (London, Tiger, 1991)

Smith, Michael, *Station X* (London, Pan Books, 1998)

Wheatley, Dennis, *The Deception Planner* (London, Hutchinson, 1980)

Whiting, Charles, *Hitler's Secret War* (London, Leo Cooper, 2000)

Journal

International Securities Studies Programme – Woodrow Wilson International Studies Centre for scholars: Andrew, Christopher M., 'The Mobilisation of British Intelligence for two World Wars' (1980)

Sources
Army Records Society
Bundesarchiv
British Library
Imperial War Museum
Intelligence Corps
King's College War Studies
Lauf Military Museum
The Royal Corps of Signals
The National Army Museum
The National Archives
The Internet

Index